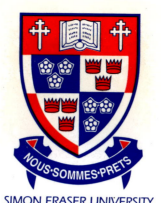

SIMON FRASER UNIVERSITY
W.A.C. BENNETT LIBRARY

Sol-Gel Materials Chemistry and Applications

ADVANCED CHEMISTRY TEXTS

A series edited by DAVID PHILLIPS, *Imperial College, London, UK*, PAUL O'BRIEN, *University of Manchester, UK* and STAN ROBERTS, *University of Liverpool, UK*

Volume 1
Chemical Aspects of Photodynamic Therapy
Raymond Bonnett

Volume 2
Transition Metal Carbonyl Cluster Chemistry
Paul J. Dyson and J. Scott McIndoe

Volume 3
Nucleoside Mimetics: Their Chemistry and Biological Properties
Claire Simons

Volume 4
Sol-Gel Materials: Chemistry and Applications
John D. Wright and Nico A.J.M. Sommerdijk

This book is part of a series. The publisher will accept continuation orders which may be cancelled at any time and which provide for automatic billing and shipping of each title in the series upon publication. Please write for details.

Sol-Gel Materials Chemistry and Applications

John D. Wright
University of Kent, UK

and

Nico A.J.M. Sommerdijk
Eindhoven University of Technology
The Netherlands

CRC PRESS

Boca Raton   London   New York   Washington, D.C.

Visit the CRC Press Web site at www.crcpress.com

© 2001 by OPA (Overseas Publishers Association) N.V.

No claim to original U.S. Government works
International Standard Book Number 90-5699-326-7
Printed in the United States of America   2 3 4 5 6 7 8 9 0
Printed on acid-free paper

Contents

Preface   ix

Chapter 1   Introduction   **1**

1.1   Colloid Stability   1
1.2   Control of Particle Nucleation   3
1.3   The Silicon Alkoxide Sol-Gel Process   3
1.4   Advantages of Sol-Gel Synthesis   4
1.5   Limitations of Sol-Gel Synthesis   5
1.6   Overall Structure of this Book   6
1.7   The Historical Development of Sol-Gel Processing   6
      1.7.1   Origins   6
      1.7.2   The Beginning of Sol-Gel Science   7
      1.7.3   The Explosion of Sol-Gel Technology   8
References   14

Chapter 2   Silica Sol-Gels: Reaction Mechanisms   **15**

2.1   Hydrolysis   15
      2.1.1   Acid and Base Catalysis   15
      2.1.2   Precursor Substituent Effects   16
      2.1.3   Hydrophobic Effects and Co-Solvents   17
      2.1.4   Effect of Water: Alkoxide Ratio (R)   18
2.2   Condensation   19
      2.2.1   Acid and Base Catalysis   19
      2.2.2   Precursor Substituent Effects   20
2.3   Overall Kinetics of Hydrolysis and Condensation   20
2.4   Non-Hydrolytic Sol-Gel Processing   21
2.5   Gelation   22
      2.5.1   Models of Gelation   23
2.6   Ageing   24
      2.6.1   Cross-Linking and Syneresis   24
      2.6.2   Coarsening and Ripening   24
      2.6.3   Phase Transformations   25
      2.6.4   The Significance of Ageing   25
2.7   Drying   26
      2.7.1   The Constant Rate Period   26
      2.7.2   The Critical Point   26
      2.7.3   First Falling-Rate Period   26

2.7.4   Second Falling-Rate Period   27
2.7.5   Consequences of Drying   27
2.7.6   Avoiding Cracking   27
2.8   Densification   28
2.8.1   Stages of Densification   28
2.8.2   Effects of Heating Rate and Gas Evolution   30
2.9   Conclusion   31
References   31

Chapter 3   The Chemistry of Sol-Gel Silicates   33

3.1   Introduction   33
3.2   Chemical Control of the Sol-Gel Process   33
3.2.1   The Hydrolysis and Condensation Reaction   33
3.2.2   Gelation   36
3.2.3   Ageing   36
3.3   Additives for Structuring and Processing   37
3.3.1   Drying Control Additives   37
3.3.2   Organic Templates   38
3.4   Ormosils   41
3.4.1   Modified Precursors   41
3.4.2   Entrapment of Functional Materials   43
3.5   Hybrid Materials   46
3.6   Surface Modification   48
3.6.1   Dehydroxylation   48
3.6.2   Surface Functionalisation   50
References   51

Chapter 4   Metal Oxide Gels   53

4.1   Introduction   53
4.2   Hydrolysis and Condensation Reactions of Metal Salt
        Precursors   54
4.2.1   The Partial Charge Model   54
4.2.2   Equilibrium Species in Aqueous Solutions of
          Metal Salts   55
4.2.3   Condensation and Polymerisation in Aqueous
          Solutions of Metal Salts   56
4.3   Effects of the Counter-Ion   59
4.4   Reactions of Metal Alkoxide Precursors   60
4.4.1   Hydrolysis and Condensation   60
4.4.2   Acid and Base Catalysis   61
4.4.3   Steric Factors and Solvent Effects   61
4.4.4   Control of Metal Alkoxide Reactions   62
4.5   The Non-Hydrolytic Sol-Gel Method for Metal Oxides   63

4.6   Particle Growth and Aggregation   64
4.7   Preparation of Monodisperse Particles   66
References   67

Chapter 5   The Characterisation of Sol-Gel Materials   69

5.1   Introduction   69
5.2   Chemical Characterisation   69
       5.2.1   Nuclear Magnetic Resonance   69
       5.2.2   Vibrational Spectroscopy   71
5.3   Physical Characterisation   72
       5.3.1   Nitrogen Adsorption Porosimetry   72
       5.3.2   Mercury Porosimetry   74
       5.3.3   Thermoporosimetry   74
       5.3.4   NMR Spin-Spin Relaxation Measurements   76
       5.3.5   NMR Spin-Lattice Relaxation Measurements   76
       5.3.6   Small-Angle Scattering   77
       5.3.7   Other Structural Techniques   80
5.4   Indirect Characterisation Methods   82
References   83

Chapter 6   Applications of Sol-Gel Silicates   85

6.1   Introduction   85
6.2   Optical Materials   85
       6.2.1   Non-Doped Glasses   85
       6.2.2   Doped Glasses   87
       6.2.3   Contact Lenses   89
6.3   Chemical Sensors   89
       6.3.1   Optical Chemical Sensing   89
       6.3.2   Biosensors   90
6.4   Catalysts   91
6.5   Coatings   93
6.6   Membranes   94
References   94

Chapter 7   Applications of Metal Oxide Sol-Gels   97

7.1   Introduction   97
7.2   Ceramics   97
7.3   Catalysts   99
7.4   Electronic Materials   101
       7.4.1   Ferroelectric Materials   101
       7.4.2   Electrochromic Materials   102
       7.4.3   Solid Electrolytes   103
       7.4.4   Other Electronic Materials   103

7.5   Flammable Gas Sensors   104
References   106

Chapter 8   The Future   **109**

8.1   The Players   109
8.2   Market Prospects   109
8.3   Developments in Characterisation Methods   110
8.4   Composite Materials   110
8.5   New Precursors   111
8.6   New Processing Methods   111
8.7   Sol-Gel Supramolecular Chemistry   112
References   114

Index   115

Chemical Substance Index   121

Preface

Sol-gel processing methods were first used historically for decorative and constructional materials. In the last century many new applications were developed, initially largely empirically but later on a more scientific basis as new characterisation techniques became available. Today sol-gel methods are reaching their full potential, enabling the preparation of new generations of advanced materials not easily accessible by other methods yet using mild, low-energy conditions. It is therefore appropriate that the topic should increasingly be included in advanced undergraduate, MSc and taught PhD courses in the areas of chemistry, physics and materials science. There is currently no concise introductory text which covers all the major areas of the subject. The aim of our book is to fill this evident gap in the market and to facilitate the development of new courses. It has been written to guide those who wish to join the growing ranks of sol-gel scientists, by providing an accessible introduction to the development, mechanisms, chemistry, characterisation methods and applications of the technique. It provides the reader with an extensive yet concise grounding in the theory of each area of the subject, as well as detailing the real and potential applications and the future prospects of sol-gel chemistry.

The task of summarising such a vast and growing multi-disciplinary field into a volume of this size and price has proved demanding, yet we believe we have gone a long way towards achieving our aim of bridging the gap between an accessible textbook and a useful research resource. The basic ideas are described clearly for the newcomer, while their development to current research level is exemplified with numerous references. The references cited, while including many illustrative original papers as well as key reviews, are intended to facilitate further exploration of the literature rather than serving as a comprehensive bibliography. We apologise to authors of many excellent papers whose work could not be included: omissions are inevitable but not deliberate.

Where possible, key fundamental ideas and other important sections of the text have been identified by shading the relevant sections. In some parts of the text, notably the chapters covering characterisation and applications, where it has been difficult to identify some sections as more important than others, little use has been made of the shading. In these areas, the absence of shaded areas should not be taken as an indication that the material is of lesser significance.

We acknowledge a considerable debt to the international sol-gel community for assisting our own development in this area. In particular the conference series of International Workshops on Glasses and Ceramics from Gels has provided personal contacts and broad perspectives. In several countries the community has set up national sol-gel groups which provide valuable support and stimulus, especially for new workers in the field. The classic text *Sol-gel Science: The Physics and Chemistry of Sol-gel Processing* by C.J. Brinker and G.W. Scherer (Academic Press: London 1990) remains an essential background reference, although the present work is designed to appeal to a different audience as explained above, and is more concise as well as including many new developments that have occurred in the decade between the two works.

Finally, we thank the publishers for their forbearance and support, and the advisers who commented in a helpful and constructive way on our manuscript.

CHAPTER 1

INTRODUCTION

Sol-gel materials encompass a wide range of inorganic and organic/inorganic composite materials which share a common preparation strategy. They are prepared via sol-gel processing involving the generation of colloidal suspensions ("sols") which are subsequently converted to viscous gels and thence to solid materials.[1] This controlled method has many advantages, which led to its historical use before the underlying scientific principles were understood. In recent years increased understanding of these principles has led to a great increase of interest in the method, and to its application in the production of a wide variety of advanced materials. To appreciate these developments it is first necessary to consider some general features of the process.

1.1   COLLOID STABILITY

A sol is a dispersion of colloidal particles[2] suspended in Brownian motion within a fluid matrix. Colloids are suspensions of particles of linear dimensions between 1 nm (10Å) and 1 μm (10^4Å). The formation of uniform suspensions of colloidal particles can be understood by calculation of the sedimentation rates assuming that the particles are spherical so that Stokes' Law may be applied. Equating gravitational and frictional forces:

$$\text{Sedimentation rate } dx/dt = [(4\pi r^3/3)(\rho' - \rho)g]/6\pi r\eta$$
$$= [2r^2(\rho' - \rho)g]/9\eta \qquad (1.1)$$

where η = viscosity of surrounding medium
   ρ = density of surrounding medium
   ρ' = density of colloid particle material
   r = radius of colloid particle.

For a material of density 2 g/ml in water, the calculated sedimentation rates are:

Radius	Sedimentation rate
10^{-9}m	2×10^{-12} m/s (8nm/h)
10^{-8}m	2×10^{-10}m/s
10^{-7}m	2×10^{-8}m/s
10^{-6}m	2×10^{-6}m/s (8mm/h)

At normal temperatures thermal motion and convection currents are sufficient to counteract any tendency for sedimentation to occur at such low rates, and uniform suspensions are observed so long as the colloid particles remain stable.

The stability of colloidal particles is determined by their resistance to aggregation, and can be remarkably high. Thus gold sols are still in existence at the Royal Institution in London which were prepared there by Michael Faraday some 150 years ago. Clearly, if all sols displayed such stability the sol-gel method would not be useful for preparing solid materials.

At first sight the stability of small colloid particles is surprising, since surface tension leads to very high pressure differences across surfaces with small radii of curvature. For a particle of radius r, density ρ and relative molar mass M, with surface tension γ, the pressure difference across the curved surface, p_r, compared to that across a flat surface, p_o, is given by the Kelvin equation:

$$RT \ln (p_r/p_o) = 2\gamma M/\rho r \qquad (1.2)$$

This has been verified experimentally, and predicts the following ratios:

r/m	p_r/p_o
10^{-7}	1.01
10^{-8}	1.1
10^{-9}	3.0

Thus small particles should tend to dissolve while larger particles should grow, as observed in Ostwald ripening of precipitates. Furthermore, attempts to generate colloidal particles by grinding solid materials frequently fail because the particles re-join under the mechanical stresses or because of attractive forces between particles, unless precautions are taken to prevent this (e.g. by grinding in presence of a surface active material, for example grind sulphur and glucose and disperse in water.) In stable sols, this is prevented because in practice colloidal particles tend to acquire surface charge by ionisation or by adsorption of ions or polar molecules from solution. The charged surface layer in turn attracts a second more diffuse layer of ions of opposite charge in the surrounding solution. The van der Waals attractive forces which potentially lead to aggregation fall off as r^{-6}. However the electrostatic repulsions between the like-charged ions of the diffuse layers around neighbouring charged colloid particles vary as r^{-1}. Thus, unless the thickness of the diffuse layer can be reduced substantially, the repulsions dominate the van der Waals attractions and the particles are stable against aggregation. Increasing the ionic strength of the solution, and in particular increasing the charge on the counter-ion, is the main way in which the diffuse layer thickness can be reduced.

For example, colloidal material in freshwater rivers is frequently precipitated at the saline limit where the river meets the seawater, leading to the formation of typical estuary features. Similarly, Al^{3+} is often used to coagulate colloidal impurities in water treatment plants. In this case, physical entrapment of the colloid particles in gelatinous $Al(OH)_3$ at high pH is a contributing factor in addition to the ionic charge effect. Conversely, if salt water floods agricultural soil, the surface Ca^{2+} ions which stabilise soil colloid particles are replaced by Na^+ which is less strongly held and easily washed off, leading to the coagulation of the soil colloid structure and formation of hard intractable masses. Treatment with gypsum (calcium sulphate) restores the original surface charge leading to eventual recovery of the soil structure.

1.2 CONTROL OF PARTICLE NUCLEATION

These environmental examples show that it is often possible to *control the physical aggregation of colloidal particles.* However, the degree of control possible in sol-gel synthesis of materials also includes:

a) *the ability to determine the sizes of the initial colloid particles,*
b) *the ways in which chemical links are formed between different colloid particles, and*
c) *the subsequent development, drying and densification of the resulting aggregates.*

Control of the sizes of initial colloid particles depends on the fact that precipitation involves 2 steps: nucleation and growth. To obtain colloids, the nucleation rate should be much faster than the growth rate. Nucleation depends on the degree of supersaturation which can be achieved before precipitation, which is determined by solubility. Thus nucleation rates will be highest for substances with very low solubility. The growth rate of particles formed by the initial nucleation depends on:

i) the amount of material available,
ii) the diffusion rate of material from solution to growing particle (limited by viscosity),
iii) the ease of orientation and incorporation of the molecules from solution into the solid lattice and
iv) the growth-inhibiting effects of impurities and other species adsorbed onto the particle surfaces.

Thus, for sparingly soluble materials the nucleation rate is very high while the amount of material available for growth of the nuclei is very small, so small particles are obtained (e.g. ferric chloride treated with boiling water gives a ferric oxide colloid — and as expected from the above discussion of colloid stability, the largest particles are obtained when the pH is close to the isoelectric point at which the net charge on the hydrolysed species is zero.[3,4]

1.3 THE SILICON ALKOXIDE SOL-GEL PROCESS

Control of the ways in which chemical links are formed between different colloid particles, and the subsequent development, drying and densification of the resulting aggregates will be illustrated in detail in chapters 2 and 3 by reference to silica produced by the sol-gel route starting with hydrolysis of silicon alkoxides. The stages in this process occur slowly enough to allow detailed study by a variety of methods such as NMR and infra-red spectroscopy. For other alkoxides, such as metal alkoxides which are much more reactive, the same series of steps, although occurring in principle, may not all be significant since some occur so rapidly that they are not amenable to either study or control. These steps are as follows:

Hydrolysis: $Si(OR)_4 + nH_2O \longrightarrow Si(OR)_{4-n}(OH)_n + nROH$

Condensation: $X_3SiOH + HOSiX'_3 \longrightarrow X_3Si\text{-}O\text{-}SiX'_3 + H_2O$ or
$X_3SiOR + HOSiX'_3 \longrightarrow X_3Si\text{-}O\text{-}SiX'_3 + ROH$

Gelation: Formation of a "spanning cluster" across the vessel, giving a network which entraps the remaining solution, with high viscosity.

Ageing: A range of processes, including formation of further cross-links, associated shrinkage of the gel as covalent links replace non-bonded contacts, Ostwald ripening and structural evolution with changes in pore sizes and pore wall strengths.

Drying: The loss of water, alcohol and other volatile components, first as syneresis (expulsion of the liquid as the gel shrinks), then as evaporation of liquid from within the pore structure with associated development of capillary stress which frequently leads to cracking. This may also include supercritical drying, in which capillary stress is avoided by the use of supercritical fluids (e.g. CO_2) in conditions where there are no liquid/vapour interfaces.

Densification: Thermal treatment leading to collapse of the open structure and formation of a dense ceramic.

1.4 ADVANTAGES OF SOL-GEL SYNTHESIS

From the above introduction, a number of reasons for the particular value of and interest in sol-gel synthesised materials become apparent:

1. The temperatures required for all stages apart from densification are low, frequently close to room temperature. Thus thermal degradation of both the material itself and any entrapped species is minimised, and high purity and stoichiometry can be achieved.
2. Precursors such as metal alkoxides and mixed alkyl/alkoxides are frequently volatile and easily purified to very high levels (e.g. by distillation or sublimation) using techniques developed for the microelectronics industry. This further contributes to high-purity products.

3. Since organometallic precursors involving different metals are frequently miscible, homogeneous controlled doping is easy to achieve.

4. The chemical conditions are mild. Hydrolysis and condensation are catalysed by acids and bases, but extreme pH conditions may easily be avoided, especially by the use of "two step" methods in which acid catalysed hydrolysis is followed by rapid neutralisation or buffering. In this way pH sensitive organic species (e.g. dyes) and even biological species including enzymes and whole cells may be entrapped and still retain their functions.

5. Highly porous materials and nanocrystalline materials may be prepared in this way.

6. By appropriate chemical modification of the precursors, control may be achieved over the rates of hydrolysis and condensation, and over colloid particle size and the pore size, porosity and pore wall chemistry of the final material.

7. Using functionalised precursors, covalent attachment of organic and biological species to porous silicate glass structures is possible.

8. By controlling the ageing and drying conditions, further pore size and mechanical strength control may be achieved.

9. By using organometallic precursors containing polymerisable organic ligands, materials may be produced which contain both inorganic and organic polymer networks.

10. Entrapped organic species may serve as templates for creation of pores with controlled size and shape. Subsequent removal of these species (for example by heat or strong acid treatment) leaves "molecular footprints" with potential as catalytic sites.

11. Since liquid precursors are used it is possible to cast ceramic materials in a range of complex shapes and to produce thin films or fibres as well as monoliths, without the need for machining or melting.

12. The optical quality of the materials is often good, leading to applications for optical components.

13. The low temperature of sol-gel processes is generally below the crystallisation temperature for oxide materials, and this allows the production of unusual amorphous materials.

1.5   LIMITATIONS OF SOL-GEL SYNTHESIS

Despite all these advantages, sol-gel materials are not without limitations. The precursors are often expensive and sensitive to moisture, limiting large scale production plants to specialised applications such as optical coatings. The process is also time-consuming, particularly where careful ageing and drying are required. Although this need not be a limiting factor where long continuous production runs are envisaged, it does mean that the total volume of material in the processing line is inevitably significantly higher than in faster processes. Finally the problems of dimensional change on densification, and of shrinkage and stress cracking on drying, although not insuperable, do require careful attention. These significant limitations emphasise the need to optimise sol-gel materials to exploit their advantages to the maximum in applications where they can provide properties not attainable by other methods.

1.6 OVERALL STRUCTURE OF THIS BOOK

The ways in which these advantages may be realised will be made clear in subsequent chapters of this book. Chapters 2 and 3 explore the reaction mechanisms and opportunity for chemical control in silica sol-gel materials, while chapter 4 develops these themes for metal oxide gels. Chapter 5 covers the methods which are available for characterisation of sol-gel materials. Chapters 6 and 7 describe the applications of silica and organically modified silica ("ormosil") sol-gels, and metal oxide sol-gels, respectively. Finally in chapter 8 we consider future prospects for sol-gel materials, with reference to developments in their chemistry, characterisation methods and applications.

1.7 THE HISTORICAL DEVELOPMENT OF SOL-GEL PROCESSING

To end this chapter we return to its first paragraph, where we pointed out that because of the advantages of sol-gel methods they were put to practical use long before any scientific understanding had been developed. A brief survey of the history of sol-gel technology[5] not only recognises the intuitive skills of earlier generations, but also serves to emphasise that despite its present high-technology niche it has its roots in constructional, decorative and artistic functions, albeit with the common objective of achieving effects in materials which could not have been achieved by other available methods of the time.

1.7.1 Origins

The earliest use of colloids to prepare functional materials is seen in the cave paintings at Lascaux in France, dating back 17,000 years. The pigments used were based on iron oxide, carbon and clays, ground into fine powders, graded by sedimentation and dispersed in water using natural oils as surface active stabilisers. It is interesting that this decorative use of a technology closely related to much more recent ceramic decoration methods substantially pre-dates the use of ceramics in construction or other more practical applications.

The next major development was the use of firing techniques in addition to simple grinding to alter the chemistry of the mineral precursors. Some 8000 years ago, early examples of the use of plaster and brick occurred. Genesis 11 v.3 describes bricks produced using "thorough burning" and held together with bitumen for the building of the tower of Babel. Neolithic sites often contain polished plaster floors. At Yiftah El in Israel a 180m^2 example would have needed over 2 tons of lime, which would have required over 10 tons of wood to fire the kiln.[6] In these examples we see the beginnings of the sol-gel ideas, in that fine powdered or colloidal material in suspension was moulded and then dried and densified by chemical action or by firing at high temperature.

Next, glazing methods were developed to seal the surfaces of porous clay vessels.[7] In China by 2000BC silicate glazes with high calcium content fired at high temperatures were in use. In Mesopotamia by the second century the glazing chemistry had already become quite sophisticated; a very fine haematite-rich illitic clay fraction made by sedimentation was painted onto the vessel and fired in a reducing atmosphere to reduce the iron to black magnetite. After sintering of the glaze, the atmosphere was made oxidising by admission of air, oxidising the iron in the non-glazed areas to a red colour and producing a shiny black glazed decorative pattern on a matt red base.

Even earlier (4000BC) in Egypt an aqueous paste of crushed sand and a sodium salt flux and binder was moulded and fired to produce *faience* by binding the silica particles together with the molten flux. Addition of copper salts to the paste led to migration of these mobile salts to the surface during firing, with formation of a translucent blue glaze.

The idea of using chemically-linked particles as a matrix for a composite of other particulate materials led to the development of concrete in about 700BC, for example in the 260m aqueduct bridge at Jerwan in Iraq. Stones of graded sizes mixed with sand and a binding paste of quicklime (CaO) and water formed the first concrete, but required an asphalt lining as the lime cement tended to crumble in water. The Romans developed concrete between 200BC and 400AD and discovered that addition of the volcanic ash from Pozzuoli near Mount Vesuvius led to a much stronger concrete which would also set under water. In fact the ash contained oxides of aluminium, silicon and iron and the product was similar to modern Portland cement, proving more durable in underwater conditions than recently-laid modern concrete in the same locations. Much the same material was rediscovered by the British engineer John Smeaton in his search for a reliable hydraulic cement for building the fourth Eddystone lighthouse[8] in the English Channel in 1756–9. He found that the best hydraulic cement was produced by burning limestone with some clay content. His work was continued by Louis Joseph Vicat[9] who examined thousands of different limestones in France for 30 years, and by Johann Friedrich John in Germany.

1.7.2 The Beginning of Sol-gel Science

Strangely, many of these early technologies became lost in the Dark Ages after the decline of the Roman Empire, and probably the next significant development was the discovery of "water glass" by von Helmont[10] in 1644. He dissolved silicate materials (stones, sand, flint etc.) in alkali and found that on acidification a precipitate of silica equal in weight to the original silicate materials was obtained. In 1779 Bergman[11] reported that if the correct amount of dilute acid was used the mixture gelled on acidification. This preparation of a silica gel led to a series of applications remarkably similar to those of today's sol-gel chemistry. In 1840–1860 it was shown to be useful as a glazing solution, as a binder for ceramic precursor powders to make porcelain and special bricks, as a mixture with sand for improving glass-making, as a method for impregnating and hardening soft porous stone, and for making "synthetic stone" using a patented process with water glass, lime or chalk, and sand. In a related development, solutions of magnesium hexafluorosilicate were used (and are still used today) for polishing marble floors, in a hydrolytic process leading to scratch resistant amorphous silica layers containing fine crystals of calcium and magnesium fluorides.

Also in the 19th century many oxide materials were prepared from hydroxide gels. Following the preparation of zirconia by firing zirconium hydroxide gels[12] (Vauqelin, 1797) Berthier[13] (1832) used mixed cupric and zirconyl salts treated with ammonia and then fired to produce a green copper zirconate, in 1842 Ebelmen[14] reported the synthesis of uranium oxide by heating the hydroxide, and in 1892 von Chroustchoff[15] heated mixed gels of zirconium hydroxide, silica and alumina in a sealed tube to obtain a zirconopyrophyllite. Meanwhile, two major developments were occurring which were to prove foundation stones for sol-gel processing: the physical properties of colloids came under intensive study by such giants as Becquerel, Faraday, Tyndall, Graham and Schulze; and in 1846 Ebelmen[16] prepared the first silicon alkoxides by the reaction between silicon tetrachloride and alcohol, observing that the product gelled on prolonged exposure to atmospheres with normal humidity. In 1876 Troost and Hautefeuille[17] made hydrolysed derivatives of silicon alkoxides. In 1884 Grimaux[18] hydrolysed tetramethoxysilane to prepare silicic acid sols, and made colloidal iron oxides from iron alkoxides. However, for the next 50 years these developments had little scientific impact for the development of the sol-gel materials field.

1.7.3 The Explosion of Sol-gel Technology

The one notable exception in this period was W.A. Patrick, who pioneered the field of silica gel desiccants, catalysts and absorbent materials, starting with the drying and firing of a homogeneous silica gel at up to 700°C to produce a very porous form of silica.[19] In 1923 he went on to show that impregnation of the partially dried materials with metal salts led to the formation of supported catalysts,[20] and by 1930 he had filed many patents[21] for supported catalysts, including the use of sol-gel methods.

This signalled the beginning of an intensive period of technological development using sol-gel methods, which produced a very large number of patents, many useful materials, and a large body of data and experience. Only later, with the advent of modern characterisation methods, could a set of firmly-based theoretical principles begin to be established to interpret all these data. Indeed it could be said that despite the new methods now available, the subject is still strongly influenced by this empirical approach. It is therefore important that the large amount of early work in the field is not overlooked, only to be re-discovered as the topics attract interest for new reasons, just as was the case with the Roman cement. Some of this work is now summarised with references:

1.7.3.1 Catalyst Materials

1925   H.N. Holmes and J.A. Anderson (Mixed gels of Al, Fe, Ca, Co, Cu, Ni and Si oxides), *J. Ind. Eng. Chem.*, 17, 280 (1925).

1945   M.M. Marisic, (Sol-gel manufacture of smooth hard catalyst support pellets of silica/alumina, alumina and silica/thoria) "Gel pellets" U.S. Patent 2,385,217 (1945). M.M. Marisic and E.M. Griest, (Thermal stability of catalyst increased by combination of zirconia gel with boria, alumina or silica gels) "Preparation of zirconia gel" U.S. Patent 2,467,089 (1949).

1945   J.A. Anderson and V. Voorhees, Use of silica/alumina sol mixture for: "Process of making spheroidal gel particles" U.S. Patent 2,468,857 (1945).

1947   G.C. Connolly, "Titania gel-boria catalyst and its method of preparation" U.S. Patent 2,424,152 (1947); G.C. Connolly, "Preparation of silica-alumina gel" U.S. Patent 2,474,888 (1949).

1947   J.G. Fife, (The versatility of sol-gel methods for catalyst production) "Catalytic Processes" British Patent 586,945 (1947).

1948   C.L. Thomas and E.C. Lee, "Preparation of a silica-alumina-zirconia catalyst" U.S. Patent 2,439,994 (1948).

1948   Development of the use of alkoxide precursors for high-purity catalysts: G. Feachem and H.T.S. Swallow, *J. Chem. Soc.*, **267** (1948); H. Atkins and S. Watkins, *J. Am. Chem. Soc.*, **73**, 2184 (1951); A.C. Coates and L. Saunders, "Improvements in the manufacture of aluminum alkoxides" British Patent 654,408 (1951); C.N. Kimberlin, Jr., "Preparation of alumina from higher alcoholates of aluminum" U.S. Patent 2,636,865 (1953); E.A. Hunter and C.N. Kimberlin, Jr., "Method for making alumina hydrosols" U.S. Patent 2,656,321 (1953); S.E. Tung and E. McIninch, *J. Catal.*, **3**, 229 (1964).

1949   J.A. Pierce and C.N. Kimberlin, Jr., (Use of B, Si, Fe, Cr, Co, Ni, Mo, Al, Ti, W oxide hydrogels in) "Preparation of spherical gel particles" U.S. Patent 2,474,910 (1949).

1949   F. Kainer, Review of alumina/silica catalysts studies: *Kolloid Zeitschrift*, **114**, 54 (1949).

1.7.3.2 Sol-gel Metal Oxides for Ceramics

1931   G. King, (Methods for hydrolysis of alkoxysilanes for use in paints, dental cements, treatment of brick, stone and concrete.) *Paint Man.*, **1**, 16–20, 52–55 (1931).

1942   D. McLachlan, Jr. (Preparation of finely divided oxides by dispersion of alkoxides or chlorides in inert media and emulsification with water.) "Method of preparing finely comminuted oxides" U.S. Patent 2,269,059 (1942).

1946   H.D. Cogan and C.A. Setterstrom (Preparation and use of partially hydrolysed tetraethoxysilane solutions.) *Chem. Eng. News*, **24**, 2499 (1946).

1949   G.A. Bole (Gelation of a ceramic powder dispersion in a precursor, within a mould, followed by drying and firing to produce a shaped ceramic article.) "Ceramic Process" U.S. Patent 2,603,570 (1952).

1948/9 A. Kreshkov and A. Vladimirov (Hydrolysis of alkoxysilanes with suspended finely divided oxides and hydroxides, for ceramic production.) A. Kreshkov, *Reports of the Academy of Sciences*, **59**, 723 (1948); A. Kreshkov and A. Vladimirov, *Reports of the Academy of Sciences*, **65**(2), 185 (1949).

1953   C.N. Kimberlin, Jr., (Preparation of 0.05–0.1μm spherical alumina particles by burning vapourised Al alkoxides in hydrogen.) "Preparation of alumina from higher alcoholates of aluminum" U.S. Patent 2,636,865 (1953); "Preparation of alumina by burning" U.S. Patent 2,754,176 (1956).

1952   (Use of homogeneous solutions of mixed alkoxides) D. Roy, thesis, Pennsylvania State College, University Park (1952); R. Roy and E.F. Osborn, *Trans. Brit. Ceram. Soc.*, **53**(9) 525 (1954); R.C. DeVries and R. Roy, *J. Amer. Ceram. Soc.*, **38**, 142 (1955); R. Roy, *J. Amer. Ceram. Soc.*, **39**(4), 145 (1956); D.M. Roy and R. Roy, *Am. Mineral.*, **40**, 147 (1955) (Metal salts + alkoxysilanes).

1954   R. Roy and E.F. Osborn, Metal organics as ceramic precursors: *Am. Mineral.*, **39**, 853 (1954).

1958   R.R. West and T.J. Gray, Formation of Mullite from metal-organics: *J. Am. Ceram. Soc.*, **41**(4), 132 (1958).

1961   G.B. Alexander and J. Bugosh, "Concentrated zirconia and hafnia aquasols and their preparation" U.S. Patent 2,984,628 (1961); A. Clearfield, "Process for the production of cubic crystalline zirconia" U.S. Patent 3,334,962 (1967); J.L. Woodhead, "Zirconium compounds" U.S. Patent 3,518,050 (1970).

1962   P. Arthur, Jr., "Process for the production of fibrous alumina monohydrate" U.S. Patent 3,056,747 (1962); O. B. Willcox, "Aluminum oxide production" U.S. Patent 3,039,849 (1962); J. Bugosh, "Fibrous alumina monohydrate and its production" U.S. Patent 2,915,475 (1959); "Chemically modifed alumina monohydrate, dispersions thereof and processes for their preparation" U.S. Patent 3,031,418 (1962); "Colloidal, anisodiametric transition aluminas and processes for making them" U.S. Patent 3,106,888 (1963).

1964   (Use of the alkoxide pyrolysis technique to prepare powders of a wider variety of metal oxides for use in oxide coatings.) K.S. Mazdiyasni and C.T. Lynch in "Special Ceramics 1964", edited by P. Popper, p 115. Academic Press: New York, 1965.

1964   (Use of sol-gel methods for nuclear fuel manufacture.) D.E. Ferguson, O.C. Dean, and D.A. Douglas, *Proc. 3rd UN Int. Conf. on Peaceful Uses of Atomic Energy*, 237 (1964); J.L. Kelly, A.T. Kleinsteuber, S.D. Clinton, and O.C. Dean, *Ind. Eng. Chem. Process Des. Dev.*, 4, 212 (1965); P.A. Haas, F.G. Kitts, and H. Beutler in "Nuclear Engineering — Part XVIII, Chemical Engineering Process Symposium Series No. 80 (American Institute of Chemical Engineers, New York, 1967), p 16; C.T. Hardy in "Sol-Gel Processes For Ceramic Nuclear Fuels", p 33. IAEA: Austria, 1968.

1967   (Low temperature sintering of oxide powders from alkoxide pyrolysis.) K.S. Mazdiyasni, C.T. Lynch, and J.S. Smith II, *J. Amer. Ceramic Soc.*, 50, 533 (1967).

1969   (Pyrolysis of mixed alkoxides to produce mixed oxides.) K.S. Mazdiyasni, R.T. Dolloff, and J.S. Smith II, *J. Amer. Ceramic Soc.*, 52, 523 (1969).

1969   H. Dislich, P. Hinz, and R. Kaufmann, "Verfahren zur Herstellung von transparenten, glasigen, glasig-kristallinen oder kristallinen anorganischen Mehrkomponentenstoffen, vorzugsweise in dunnen Schichten, bei Temperaturen weit unterhalb des Schmelzpunktes" FRG-Patent 19 41191 (1969); H. Dislich, *Angewandte Chemie Int. Ed.*, 10(6), 363 (1971); H. Dislich and P. Hinz, *J. Non-Cryst. Solids*, 48, 11 (1982).

1975   R.N. Howard and H.G. Sowman, "Spheroidal composite particle and method of making" U.S. Patent 3,916,584 (1975); M.A. Leitheiser and H.G. Sowman, "Non-fused aluminum oxide-based abrasive mineral" U.S. Patent 4,314,827 (1982).

1975   B.E. Yoldas, Ceramic Bulletin, 54(3), 286 (1975) (porous, transparent monolithic alumina); B.E. Yoldas, "Transparent activated nonparticulate alumina and method of preparing same" U.S. Patent 3,941,719 (1976);

1.7.3.3 Sol-gel Films and Coatings

1928   G. King and R. Threlfall, "Material for forming coatings, for use as impregnating agents or for like purposes" U.S. Patent 1,809,755 (Appl. 1928).

1934   A.B. Ray, "Method of depositing silica on material" U.S. Patent 2,027,931 (Appl. 1934).

1938   C.P. Marsden, "Method of coating glass" (coating the interior walls of light bulbs and vacuum tubes using hydrolysed alkoxysilane solutions) U.S. Patent 2,329,632 (Appl. 1938).

1941   C.J. Christensen (Hydrolysis of an alkoxysilane with an organic filler to produce an insulating organic/inorganic composite.) "Switching Device" U.S. Patent 2,347,733 (Appl. 1941).

1942   W. Geffcken and E. Berger, "Method for producing layers on solid objects" U.S. Patent 2,366,516 (1945 [This is the U.S. equivalent of German application ser. no. 333,186 (1943)]. (A method of spraying a precursor fluid, e.g. silicon alkoxides, aluminium acetylacetonate and different metal halides, onto heated glass, precipitating a gel of hydrated oxide which could be densified at low temperature to form a scratch-resistant anti-reflection coating.) See also W. Geffcken, *Angew. Chem.*, **60**(1), 1 (1948); W. Geffcken, *Zeitschrift fur Glaskunde*, 24, 143 (1951). H. Schroeder, *Phys. Thin Films*, **5**, 87–141 (1969).

1947   American Optical Company: Use of partially hydrolysed alkoxysilanes, titanium oxychlorides, tin chloride and various colloids in mixtures for anti-reflection coatings. H.R. Moulton, "Reflection reducing coating having a gradually increasing index of refraction" U.S. Patent 2,432,484 (1947); H.R. Moulton and E.D. Tillyer, "Reflection modifying coatings and articles so coated and method of making the same" U.S. Patent 2,466,119 (1948); H.R. Moulton, "Reflection reducing coatings having uniform reflection for all wavelengths of light and method of forming such coatings" U.S. Patent 2,531,945 (1950); H.R. Moulton, "Method of forming a reflection reducing coating" U.S. Patent 2,536,764 (1951); H.R. Moulton, "Surface reflection modifying solutions" U.S. Patent 2,584,905 (1952); H.R. Moulton, "Composition for reducing the reflection of light" U.S. Patent 2,601,123 (1952).

1953   (Anti-reflection coatings using colloidal silica plus compounds of Cr, Mn, Fe, Ni, Cu, Zn, Ag, Ti, Sn and Pb.) S. Mclean "Method of forming a transparent reflection-reducing coating on glass and the article resulting therefrom" U.S. Patent 2,639,999 (1953).

1954   (Formic acid catalysed hydrolysis of alkoxysilanes for producing coatings.) E.A. Thurber "Process for producing a silica coating" U.S. Patent 2,692,838 (1954).

1959   (TiO_2-SiO_2-TiO_2 coated rear-view mirrors [Schott].) — see H. Dislich and P. Hinz, *J. Non-Cryst. Solids*, **48**, 11 (1982).

1963   H. Schroder "Transparent, conductive, reflection-reducing coatings on non-conductive objects and method" U.S. Patent 3,094,436 (1963).

1.7.3.4 Composites

1946   (Generation and purification of organosilanes from acidified silicate solutions, and their application to catalysis and polymer composites.) J.S. Kirk "Chemical processes and products" U.S. Patent 2,395,880 (1946); R.K. Iler and J. S. Kirk, "Silicic acid compositions" U.S. Patent 2,408,655 (1946); J.S. Kirk, "Process for producing silicic acid sols" U.S. Patent 2,408,656 (1946); R.K. Iler and P.S. Pinkney, *Ind. Eng. Chem.*, **39**(11), 1379 (1947).

1953   (Composite formation by gelation of fine colloidal silica with an organic emulsion.) J.H. Wills and J.F. Hazel, "Manufacture of silica filled materials" U.S. Patent 2,649,388 (1953).

1.7.3.5 Aerogels

1932   (Aerogels by replacement of water by alcohol and removal under supercritical conditions.) S.S. Kistler, *Nature*, **127**, 711 (1931); *J. Phys. Chem.*, **36**, 52 (1932); "Inorganic aerogel compositions" U.S. Patent 2,188,007 (1940); S.S. Kistler, E.A. Fischer, and I.R. Freeman, *J. Am. Chem. Soc.*, **65**, 1909 (1943).

1.7.3.6 Fibres

1960   E. Wainer, B.C. Raynes, and A.L. Cunningham, "Liquid polymers, solid articles made therefrom and methods of preparing same" U.S. Patent 3,180,741 (1960).

1963   R.M. Beasley and H.L. Johns, "Inorganic fibers and method of preparation" U.S. Patent 3,082,099 (1963); R.H. Kelsey, "Preparation of inorganic oxide monofilaments" U.S. Patent 3,311,689 (1963).

1967   J.E. Blaze, Jr., "Process of manufacturing refractory fibers" U.S. Patent 3,322,865 (1967).

1973   G. Winter, M. Mansmann, and H. Zirngibl, "Inorganic fibres and a process for their production" British Patent 1,323,229 (1973).

1973   H.G. Sowman, "Refractory fibers and other articles of zirconia and silica mixtures" U.S. Patent 3,709,706 (1973); H.G. Sowman, "Aluminum borate and aluminum borosilicate articles" U.S. Patent 3,795,524 (1974). See also H.G. Sowman, *Ceramic Bulletin*, **67**(12), 1911 (1988).

1974   L.E. Seufert, "Alumina fiber" U.S. Patent 3,808,015 (1974).

1974   W. Verbeek, "Production of shaped articles of homogeneous mixtures of silicon carbide and nitride" U.S. Patent 3,853,567 (1974).

REFERENCES

1. C.J. Brinker and G.W. Scherer, "Sol-gel science: The physics and chemistry of sol-gel processing", Academic Press: London, 1990.
2. "Colloid and Surface Chemistry" D.J. Shaw, Butterworth, 4th ed., 1992.
3. K.M. Towe and W.F. Bradley, *J. Colloid Interface Sci.*, 24, 384–392 (1967).
4. T. Sugimoto and E. Matijevic, *ibid*, 74, 227–243 (1980).
5. T.E. Wood and H. Dislich "An abbreviated history of sol-gel technology" in "Sol-gel Science and Technology", edited by E.J.A. Pope, S. Sakka and L. Klein, *Ceramic Transactions*, v. 55, American Ceramic Society, Westerville, Ohio, 1995.
6. W.H. Gourdin and W.D. Kingery, *J. Field Archaeology*, 2, 133 (1976); W.D. Kingery, P.B. Vandiver and M. Prickett, *J. Field Archaeology*, 15, 219 (1988).
7. P.B. Vandiver, *Scientific American*, 262, p. 106 (1990).
8. J. Smeaton, "Narrative of the building etc. of the Eddystone Lighthouse", London, 1791.
9. L.J. Vicat, "Recherches experimentales sur les chaux de construction, les betons, et les mortiers ordinaires", Oarusm, 1818; *Journ. Phys.*, 86, 189 (1818); *Ann. Chim. Phys.*, 5, 387 (1817).
10. J.B. von Helmont, *De lithiase, Amstelodami*, 28 (1644).
11. T. Bergman, "De terra silicea", *Upsala*, 1779.
12. L.N. Vauquelin, *Ann. Chim. Phys.*, 22, 179 (1797).
13. P. Berthier, *Ann. Chim. Phys.*, 50, 362 (1832).
14. J.J. Ebelmen, *Ann. Chim. Phys.*, 5, 199 (1842).
15. K. von Chroustchoff, *Bull. Acad. St. Petersburg*, 35, 343 (1892).
16. J.J. Ebelmen, *Compt. Rend.*, 19, 398 (1844); *Ann. Chim. Phys.*, 16, 129 (1846); *J. Pharm. Chim.*, 6, 262 (1844).
17. L. Troost and P. Hautefeuille, *Ann. Chim. Phys.*, 7, 452 (1876).
18. E. Grimaux, *Compt. Rend.*, 98, 105, 1434, 1485 (1884); *Bull. Soc. Chim.*, 41, 50, 157 (1884).
19. W.A. Patrick, "Silica gel and process of making same" U.S. Patent 1,297,724 (1919).
20. W.A. Patrick, "Catalysts" British Patent 208,656 (1923).
21. W.A. Patrick, "Adsorbent and catalytic oxides" U.S. Patent 1,520,305 (1924); "Porous catalytic gels" U.S. Patent 1,577,186 (1926); "Gels for catalytic and adsorbent purposes" U.S. Patent 1,577,189 (1926); "Gels for catalytic and adsorbent purposes" U.S. Patent 1,577,190 (1926); "Tungstic oxide gel" U.S. Patent 1,682,239 (1928); "Stannic oxide gel" U.S. Patent 1,682,240 (1928); "Aluminum oxide gel" U.S. Patent 1,682,241 (1928); "Titanium oxide gel" U.S. Patent 1,682,242 (1928); "Catalytic and adsorbent gel" U.S. Patent 1,696,644 (1928); "Catalytic and adsorbent gel" U.S. Patent 1,696,645 (1928); "Catalytic material" U.S. Patent 1,695,740 (1928); "Silica gel adsorption separations" U.S. Patent 1,537,260 (1925); "Gels; Absorbents for gases; catalytic agents" British Patent 159,508 (1921); W.A. Patrick and E.H. Barclay, "Tungsten oxide gel" U.S. Patent 1,848,266 (1932); "Tungsten oxide gel" U.S. Patent 1,683,695 (1928).

CHAPTER 2

SILICA SOL-GELS: REACTION MECHANISMS

Reaction mechanisms must be considered if sol-gel chemistry is to be developed as anything other than an empirical art. In this chapter we shall describe several mechanistic principles which allow some prediction of the effects of varying the nature and amounts of the silicon precursor, co-solvent and water, and of variables such as pH and temperature. For convenience, the stages of sol-gel processing described in chapter 1 will be considered separately, although it must be realised that this is an artificial scenario since in real systems several of the steps may occur concurrently.

2.1 HYDROLYSIS

2.1.1 Acid and Base Catalysis

The first step of the hydrolysis of a silicon alkoxide can occur by acid catalysed or base-catalysed processes, as shown in figure 1:

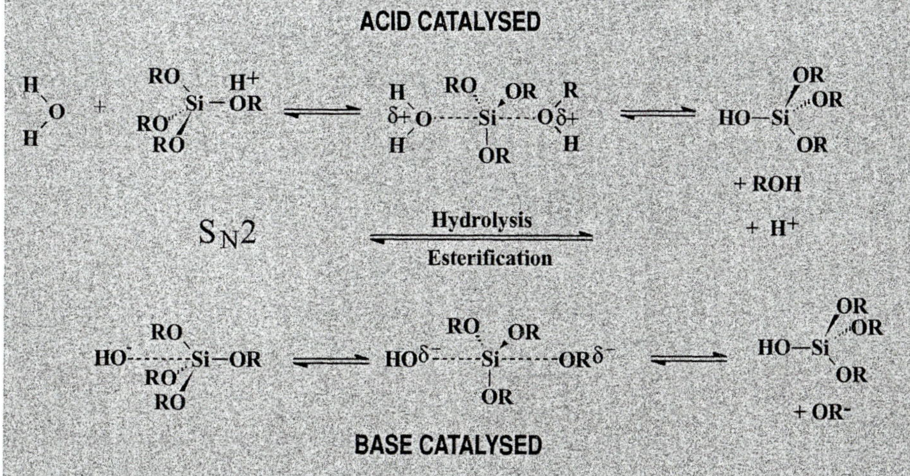

FIGURE 2.1 HYDROLYSIS MECHANISMS.

The effect of catalysis may be judged by comparing the rate of reaction at different pH values, bearing in mind that the isoelectric point of silica (where the equilibrium species has zero net charge) is at pH 2.2. Although it is only a crude measure of the relative hydrolysis rates because it includes both hydrolysis and condensation, the time to form a gel (t_{gel}) gives an indication of the relative rates, as shown in figure 2.2:

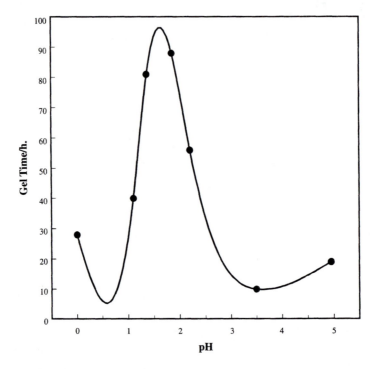

FIGURE 2.2   GEL TIME AS A FUNCTION OF pH FOR HCL-CATALYSED TEOS (H_2O:TEOS RATIO, R, = 4).[1]

As expected, the gel time is longest at the isoelectric point, and rapidly decreases in acid or base conditions relative to the isoelectric point pH. (The slight rise at high pH is due to dissolution of silica in highly alkaline conditions.)

2.1.2   Precursor Substituent Effects

The rate trends in acid and base catalysed processes for successive hydrolysis of the four alkoxy groups around silicon can be understood in terms of electronic effects. Alkoxy groups are more electron donating than hydroxy groups. Thus for the positively-charged transition state of the acid-catalysed reaction, as more alkoxy groups are replaced by hydroxy groups the transition state becomes less stabilised and the reaction rate decreases. Conversely, for the negatively-charged transition state of the base-catalysed reaction, more OH groups mean more stabilisation of the transition state and faster reaction. Similar arguments obviously show that the reverse esterification reactions are more likely in acidic conditions, where esterification stabilises the transition state, than in basic conditions.

When the rates of hydrolysis of different silicon alkoxides are compared, it is found that the steric bulk of the alkoxy group exerts a large influence. Larger alkoxy groups lead to more steric hindrance and overcrowding of the transition state, and thus lead to slower reactions. Thus tetramethoxy silane (TMOS) hydrolyses faster than tetraethoxy silane (TEOS), and data for other silicon alkoxides are shown in tables 2.1 and 2.2:[2]

TABLE 2.1

Si(OR)$_4$; R =	Hydrolysis rate constant/ 10^{-2}l.mol^{-1}s^{-1}(H$^+$)$^{-1}$
C$_2$H$_5$-	5.1
C$_4$H$_9$-	1.9
C$_6$H$_{13}$-	0.83
((CH$_3$)$_2$CH(CH$_2$)$_3$CH(CH$_3$)CH$_2$-	0.30

TABLE 2.2

Compound		Hydrolysis rate constant/ 10^{-2}l.mol^{-1}s^{-1}(H$^+$)$^{-1}$
Si(OC$_6$H$_{13}$)$_{4-n}$(OC$_2$H$_5$)$_n$	n = 0	0.83
	n = 1	1.1
	n = 2	5.0
Si(OEt)$_2$[OCHMeCH$_2$CH(CH$_3$)$_2$]$_2$		0.15
Si(OEt)$_2$[OCHMe(CH$_2$)$_5$CH$_3$]$_2$		0.095

2.1.3 Hydrophobic Effects and Co-solvents

In addition to the above-mentioned electronic and steric effects of the substituents, the hydrophobic or hydrophilic character of the precursor must also be taken into account. Because of the hydrophobic nature of the ethoxy groups, TEOS and water are immiscible in all proportions and it is necessary to add a co-solvent to achieve miscibility to facilitate hydrolysis. Figure 2.3 shows the phase diagram for TEOS/ethyl alcohol/water.

This phase diagram defines the minimum amount of alcohol required to achieve miscibility in any given mixture of TEOS and water. However, since alcohol is a product of the hydrolysis reaction, if the precursor and water can be mechanically mixed and induced to react (e.g. by use of an ultrasonic probe) eventually a single-phase mixture can be obtained. Clearly, if larger, more hydrophobic, alkyl or aryl substituents replace the ethyl groups in TEOS this problem of miscibility increases. If mixtures of two or more different precursors are used, such effects may lead to phase separation in the resulting gels. Many different co-solvents have been used, including different alcohols, formamide, dimethylformamide, 1,4-dioxane and tetrahydrofuran. The choice of added co-solvent is important, since use of a different

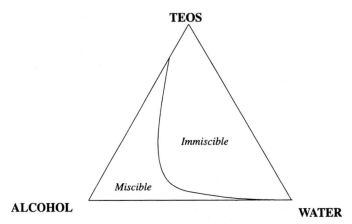

FIGURE 2.3   THE PHASE DIAGRAM FOR TEOS/ETHANOL/WATER.

alcohol from that generated by hydrolysis of the alkoxide can lead to trans-esterification and affect the whole hydrolysis and condensation reaction sequence. The co-solvent may also influence the drying process, as will be discussed later in this chapter. Where a co-solvent is chosen primarily for the latter effect it is often referred to as a drying control chemical additive (DCCA) — see also chapter 3.

2.1.4   Effect of Water: Alkoxide Ratio (R)

As seen from figure 2.3 the ratio of water:alkoxide (R) determines the amount of co-solvent required, but this ratio also influences the reaction rate. The stoichiometric ratio of water:alkoxide for complete hydrolysis is 4:

$$Si(OEt)_4 + 4H_2O \rightleftharpoons Si(OH)_4 + 4EtOH$$

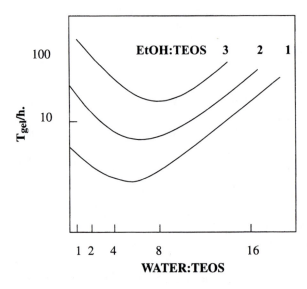

FIGURE 2.4   GEL TIMES AS A FUNCTION OF ETHANOL:TEOS AND WATER:TEOS RATIOS.[3]

However, less water than this can be used since the condensation reaction leads to production of water:

$$Si(OEt)_3OH + HOSi(OEt)_3 \rightleftharpoons (EtO)_3SiOSi(OEt)_3 + H_2O$$

If the amount of water becomes very small, however, the hydrolysis rate slows down due to the reduced reactant concentration. Similarly, if very large amounts of water are used the other reactant (alkoxide) is effectively diluted and gel times increase. The effects on gel times of varying ethanol:TEOS and water:TEOS molar ratios are shown in figure 2.4.

2.2 CONDENSATION

2.2.1 Acid and Base Catalysis

Condensation reactions can be either water condensation (as shown above) or alcohol condensation:

$$Si(OEt)_3OR + HOSi(OEt)_3 \rightleftharpoons (EtO)_3SiOSi(OEt)_3 + ROH$$

The reverse reactions are hydrolysis and alcoholysis, respectively. As with initial hydrolysis, condensation reactions may be acid or base catalysed, and in either case the reaction proceeds via a rapid formation of a charged intermediate by reaction with a proton or hydroxide ion, followed by slow attack of a second neutral silicon species on this intermediate:

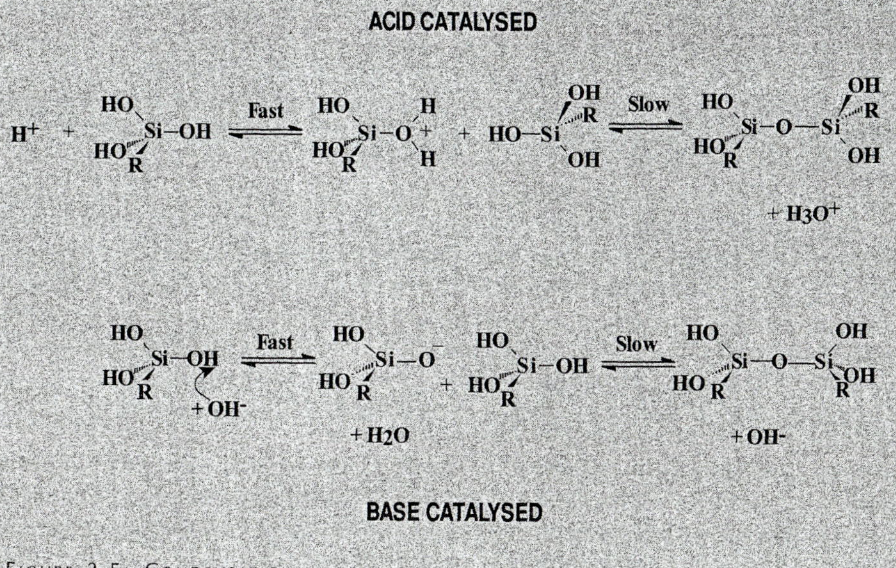

FIGURE 2.5 CONDENSATION MECHANISMS.

2.2.2 Precursor Substituent Effects

Just as with hydrolysis, the relative rates of reaction of different species depend on steric effects and the charge on the transition state. Thus for acid hydrolysis with a positively charged transition state stabilised by electron donating groups, $(RO)_3SiOH$ condenses faster than $(RO)_2Si(OH)_2$, which condenses faster than $(RO)Si(OH)_3$ etc. This means that for acid catalysed reactions, the first step of the hydrolysis is the fastest, and the product of this first step also undergoes the fastest condensation. Hence an open network structure results initially, followed by further hydrolysis and cross-condensation reactions. In contrast, in base catalysed conditions the negatively charged transition state becomes more stable as more hydroxy groups replace the electron donating alkoxy groups. Thus successive hydrolysis steps occur increasingly rapidly, and the fully hydrolysed species undergoes the fastest condensation reactions. As a consequence, in base catalysed reactions highly cross-linked large sol particles are initially obtained which eventually link to form gels with large pores between the interconnected particles. (See figure 2.6 for a diagrammatic representation.) Hence the choice of acid or base catalysis has a substantial influence on the nature of the gel which is formed (see also chapter 3).

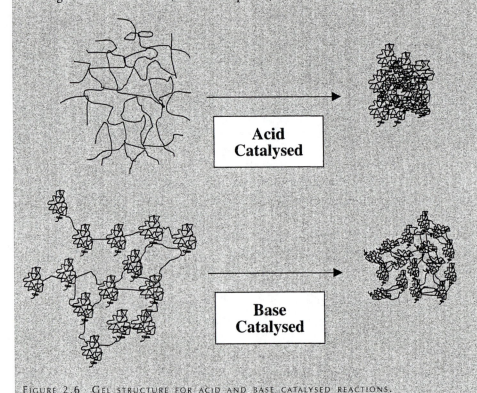

Acid Catalysed

Base Catalysed

FIGURE 2.6 GEL STRUCTURE FOR ACID AND BASE CATALYSED REACTIONS.

2.3 OVERALL KINETICS OF HYDROLYSIS AND CONDENSATION

In real systems, where hydrolysis and condensation proceed concurrently, the overall reaction kinetics rapidly become very complex. Thus, for a system SiABCD, A, B,

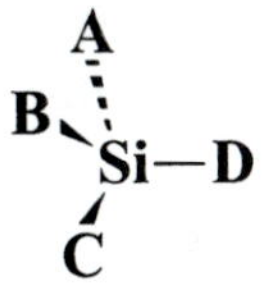

$$
\begin{array}{c}
\mathbf{A} \\
\mathbf{B} \diagdown \\
\quad \mathbf{Si{-}D} \\
\diagup \\
\mathbf{C}
\end{array}
$$

C, and D may be OR, OH or OSi, and the system is only defined in terms of $Si(OR)_x(OH)_y(OSi)_z$ with $x + y + z = 4$. Even if the diverse possibilities for the second z silicon species are ignored, this still gives 15 chemically different species, related by 10 hydrolysis reactions of the type:

$$SiOR + H_2O \rightarrow SiOH + ROH$$

55 water condensation reactions of the type:

$$2SiOH \rightarrow SiOSi + H_2O$$

and 100 alcohol condensation reactions of the type:

$$SiOH + SiOR \rightarrow SiOSi + ROH$$

(since there are in fact 10 different species each for "SiOH" and "SiOR").

Even if the reverse reactions are ignored, 165 rate constants would need to be determined to characterise the overall reaction rate fully.[4] If the next-nearest functional group environments are also considered, the number of distinct local silicon environments jumps to 1365, related by 199,290 forward-reaction rate constants. Clearly, since the formation of a gel involves extending the condensation over many more shells than this and since the reverse reactions also need consideration, it is totally impossible to develop full kinetic models for the sol-gel process. Fortunately modern high-resolution ^{29}Si NMR experiments do provide a means for determining the relative proportions of different silicon species in both liquid and gelled materials (Chapter 5), providing a rational basis for understanding the overall effects of changing the chemical and physical variables in the sol-gel process.

2.4 NON-HYDROLYTIC SOL-GEL PROCESSING

Recently an alternative non-hydrolytic sol-gel process has been developed.[5,6] This uses condensation reactions between metal halides and alkoxides:

$$M\text{-}Cl + M\text{-}OR \longrightarrow M\text{-}O\text{-}M + R\text{-}Cl$$

The alkoxide itself may be formed in-situ by reaction of the metal halide with ethers or alcohols:

$$M\text{-}Cl + R\text{-}O\text{-}R \longrightarrow M\text{-}OR + R\text{-}Cl$$
$$M\text{-}Cl + R\text{-}OH \longrightarrow M\text{-}OR + HCl$$

The reaction may be carried out in a sealed tube at temperatures typically of the order of 110°C, but will also proceed at atmospheric pressure in unsealed vessels under nitrogen.[7]

In the absence of a catalyst, these reactions only occur for silicon compounds if R is a tertiary allylic or benzylic group capable of stabilising a carbocation intermediate.[8] For example, $SiCl_4$ reacts with benzyl alcohol forming a gel after 12 hours at room temperature. The reaction products present in the syneresis liquid are benzyl chloride and HCl, showing that the coordinated benzyl alcohol intermediate reacts to give Si-OH as well as $Si-OCH_2Ph$ products, with the former undergoing condensation reactions with Si-Cl. Reactions of this type for a wide range of silicon compounds can be catalysed by small amounts of metal chlorides (e.g. 0.1–1% $FeCl_3$, $AlCl_3$, $TiCl_4$, $ZrCl_4$ etc.) which become incorporated into the product yielding coloured glasses.[8] A possible mechanism for this process involves the concerted cyclic reaction shown in figure 2.7:

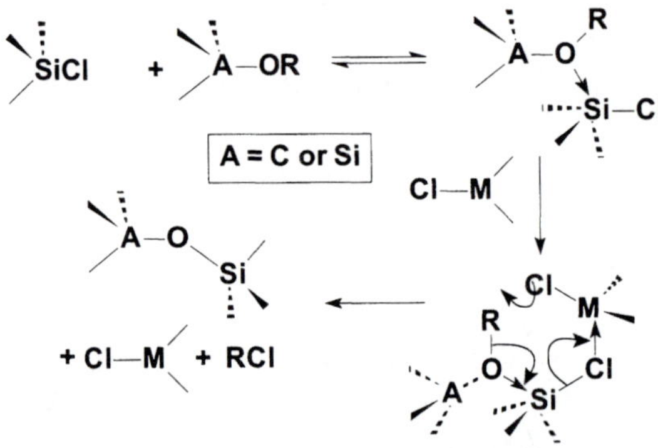

FIGURE 2.7   MECHANISM OF METAL CHLORIDE-CATALYSED NON-HYDROLYTIC SOL-GEL FORMATION OF SILICA.

The incorporation of the metal chloride in the product is due to the fact that similar reactions to those shown in figure 2.7 can also occur for the case where the metal chloride itself replaces the Si-Cl species. This means that the metal chloride is strictly not simply a catalyst, and that detailed understanding of the reaction kinetics is complicated by the evolving changes in the nature of the initial catalytic species as the reaction proceeds.

These non-hydrolytic sol-gel processing reactions are generally carried out in sealed tubes at temperatures of 110°C, with gelation times typically ranging from a few hours to a few tens of hours depending on the catalyst and the nature of the oxygen donor species. The products have low water and silanol content, high surface areas (typically 400–800m²/g) and varying texture and pore size distribution depending on the nature of the oxygen donor and catalyst.[8]

2.5   GELATION

Gelation occurs when links form between silica sol particles, produced by hydrolysis and condensation, to such an extent that a giant spanning cluster reaches across the containing vessel. At this point, although the mixture has a high viscosity so that it does not pour when the vessel is tipped, many sol particles are still present as such, entrapped and entangled in the spanning cluster. This initial gel has a high viscosity but low elasticity. There is no exotherm or endotherm, nor any discrete chemical change, at the gel point; only the sudden viscosity increase (figure 2.8). Following gelation, further cross-linking and chemical inclusion of isolated sol particles into the spanning cluster continues, leading to an increase in the elasticity of the sample. Precise definition of terms such as "gel-point" and "gelling time" with reference to attainment of a particular viscosity value is elusive and probably of little absolute significance.

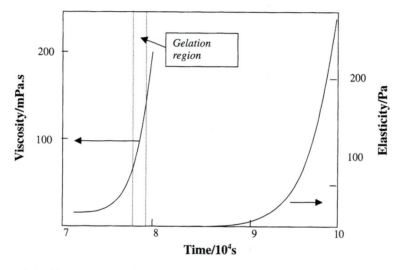

FIGURE 2.8   VISCOSITY AND ELASTICITY CHANGES IN RELATION TO GELATION.

2.5.1   Models of Gelation

The classical theory of gelation by Flory and Stockmayer[10] addresses the question of what fraction of all the possible bonds that could form in a polymerising system actually need to form before an infinitely large molecule appears. In its simple expression, considering only chain extension and branching, this approach predicts that the density of growing chains at the periphery of a growing polymer particle will increase as the particle grows. This will eventually lead to overcrowding and the prediction of a density which increases indefinitely with particle size, and is thus unsatisfactory. Percolation models of gelation are based on random filling of sites, or formation of bonds between sites, on lattices.[11] The absence of any correlation between successive bond formations in such models make it difficult to account for the formation of systems composed of sol particles using percolation models. However, the linking of sol particles to form gels can be described in this way. The latter process is also described by a kinetic model in which the rate of change in the number of clusters n_s of size s is described by the *Smoluchowski Equation*[12] in terms of their

$$dn_s / dt = \frac{1}{2} \sum_{i+j=s} K(i,j) n_i n_j - n_s \sum_{j=1}^{\infty} K(s,j) n_j \qquad (2.1)$$

formation by aggregation of clusters of sizes i and j (first term) and their loss by further aggregation (second term). Computer simulations[13] of cluster aggregation agree well with the predictions of this model. However the kinetic model gives no clues as to the detailed structure of the clusters.

2.6 AGEING

2.6.1 Cross-linking and Syneresis

Although understanding of gelation is important in applications requiring processing of either fluid (e.g. spin- or dip-coating) or self-supporting (e.g. casting) precursors, the continuing chemical and physical changes during ageing after gelation are of even more importance. NMR studies of gelled samples[14] show a continuing gradual increase in the number of Q^3 and Q^4 Si species (i.e. Si attached via 4 oxygen links to three and four other silicon atoms), due to cross-linking via trans-pore condensation reactions of pore-surface hydroxy groups. This can continue for months for samples at room temperature, the rate depending on pH, temperature and gel composition. The net effect of these processes is a stiffening and shrinkage of the sample. Shrinkage occurs because new bonds are formed where there were formerly only weak interactions between surface hydroxy and alkoxy groups. This shrinkage leads to expulsion of liquid from the pores of the gel, so that gel samples in sealed containers gradually change in appearance from homogeneous gels to transparent shrunken solid monoliths immersed in liquid. This process is known as syneresis.

2.6.2 Coarsening and Ripening

Another process associated with ageing is often referred to as coarsening or ripening. In this process, material dissolves from the surface of large particles and deposits on the initially narrow "necks" which join particles to each other. This process can best be understood by reference to figure 2.9.

The surface of an individual particle has a positive radius of curvature (r_+), whereas that of the narrow neck between particles has a smaller negative radius of curvature (r_-). As explained in chapter 1, the Kelvin equation predicts a pressure gradient across a curved interface between two media, and this leads to a difference in solubility of the material in the two regions shown in figure 2.8. The solubility S of a curved surface of radius of curvature r of a solid with interfacial tension γ_{SL} is related to the solubility of a flat surface (S_0) by the equation:

$$S = S_0 \exp(2 \gamma_{SL} V_m / RTr) \qquad (2.2)$$

where V_m is the molar volume of the solid and T is the temperature.

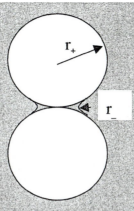

FIGURE 2.9 RADII OF CURVATURE OF PARTICLES AND "NECKS".

Thus, small particles will have higher solubility, whereas the regions of negative surface curvature will have low solubility and will tend to accumulate material. This will strengthen the solid as well as leading to some change in pore sizes and shapes. Like the continuing cross-linking processes, it will depend on temperature and pH which both influence S_0. It is also influenced by the ageing medium, which need not be water, and by other physical parameters such as pressure.

2.6.3 Phase Transformations

A final ageing effect may be phase transformation. Where gelation has occurred very quickly (e.g. in base catalysed conditions) or where several precursors of different miscibility with water have been used, there is a possibility that the porous gel contains isolated regions of unreacted precursor. On prolonged soaking in water, this material may react either completely or partially, giving inclusions of material of different structure and composition. If the refractive index of such regions is sufficiently different from that of the host matrix, the whole sample may have a white opaque appearance characteristic of a phase-separated material. This undesirable situation may be avoided by modifying the reaction rate (e.g. by dilution or pH control) and use of more effective co-solvents in the initial mixture.

2.6.4 The Significance of Ageing

Ageing effects are often cited as a significant disadvantage in the use of sol-gel materials in technological applications, particularly where the sol-gel method is proposed as a low-temperature mild-conditions method for entrapment of organic or biological species. It is therefore important to consider the following points arising from the above discussion:

i)   Ageing usually improves the properties of the material;

ii)   The ageing process can be controlled by varying the pH, temperature, pressure, ageing liquid medium and initial precursor mixture composition, and may thus be optimised;

iii) Where regular production is envisaged, ageing need not lead to production delay once an initial aged stock has been generated, provided production requirements can be anticipated.

2.7   DRYING

There are four main stages in the drying of a gelled sample:

2.7.1   The Constant Rate Period

Initially a gel will shrink by an amount equal to the volume of water or other liquid which has evaporated. This phase can only occur in gels which are still very flexible and compliant, and able to adjust to the reduced volume. (Note that gels may also shrink faster than the water can evaporate if rapid cross-linking and syneresis is occurring. If the gel has become rigid due to cross-linking by the time the excess water has evaporated, the constant rate period may cease as soon as the excess water arising from syneresis has evaporated. In this case the pore size distribution will be strongly influenced by the cross-linking, whereas gels which shrink by evaporation of water from compliant structures will suffer pore collapse as shrinkage proceeds. These effects on pore size distributions are discussed later.)

2.7.2   The Critical Point

As the gel dries and shrinks, its more compact structure and associated additional cross-linking lead to increased stiffness. At the critical point, the gel becomes sufficiently stiff to resist further shrinkage as liquid continues to evaporate. At this point the liquid begins to recede into the porous structure of the gel. Due to its surface tension and the small size of the gel pores, very large pressures are generated across the curved interfaces of the liquid menisci in the pores. For a typical silica sol-gel sample of surface area several hundred m^2/g, the forces may exceed 100 MPa (approx. 1000 atmospheres pressure!). Unless the gel has been very carefully prepared to have optimum cross-linking, as well as been very carefully aged, it will crack due to this capillary stress.

2.7.3   First Falling-Rate Period

Due to the hydrophilic nature of the pore walls and capillary forces, in many gels as the bulk of the liquid recedes into the capillary pores a thin liquid film remains on the pore walls. Flow in this film to the surface followed by evaporation, as well as direct evaporation from the filled pore region, leads to further drying. From the Kelvin equation it follows that at a given vapour pressure all menisci must have the same radius of curvature. Taken together with contact angle considerations, this means that the menisci first recede into the largest pores only. As these empty, the vapour pressure decreases and smaller pores begin to empty, with increased capillary stress. Thus, cracking may occur at any stage in this phase of the drying.

2.7.4   Second Falling-Rate Period

As the meniscus recedes from the surface, it becomes increasingly difficult to maintain the liquid film on the pore wall, due to evaporation, and at the onset of the second falling-rate period this film is broken and further liquid transport from the filled portion of the pore must involve a vapourisation step before the liquid reaches the surface. Prediction of the drying rate during this phase is difficult, as it depends on the pore-size distribution, the relative temperatures of the bulk and surface of the sample, and the possible presence of isolated pockets of liquid at irregular pore surfaces between the main liquid interface within the pores and the outer gel/air interface. For example, if the bulk of the sample is cooled as the liquid evaporates and the external surface is at a higher temperature, a reverse flow of vapour back into the bulk of the sample may even occur. Generally, however, a continuous loss of liquid at a gradually decreasing rate is observed.

2.7.5   Consequences of Drying

Several interesting observations may be explained in terms of the above stages:

i)   As a flat monolith sample dries with its lower face in contact with the containing vessel, evaporation is greater from the upper surface. The lower region of the sample therefore has filled pores while an increasingly large region of the upper surface has empty pores. The region of empty pores is able to distribute the residual capillary stress, and thus expands slightly relative to the lower region. Hence these samples tend to develop lower concave surfaces, even on flat-bottomed containing vessels.

ii) Soluble materials from the bulk of the material may be transported along the thin film of liquid on pore walls and deposited as a white efflorescence at or near the surface of the sample. This may be avoided by washing out the material before drying.

iii) As surface layers dry before the bulk of the sample, a slight surface cloudiness may sometimes be observed in partially-dried samples, depending on pore sizes.

2.7.6   Avoiding Cracking

i)   All the above problems of stress and cracking may be avoided by the use of supercritical drying. In this process water is first exchanged for alcohol, which is then removed under super-critical conditions where the distinction between liquid and vapour no longer exists. Hence no capillary stresses are set up during the drying. This requires high temperatures and pressures, and is expensive and potentially dangerous. The use of alcohol is necessary as the original research (see chapter 1) showed that silica gel would dissolve in water under the conditions necessary to obtain supercritical water. One consequence of the use of supercritical alcohol is that the gels are hydrophobic, due to re-esterification reactions of surface hydroxyl groups in supercritical alcohol. These materials are known as *aerogels* (whereas gels dried by conventional evaporation are known as *xerogels*), and have only slightly smaller volumes than those of the original wet gels, although they collapse on re-immersion in liquids. Recently progress has been made on the use of supercritical carbon dioxide in this process, although complete preliminary solvent exchange is necessary as carbon dioxide is immiscible with water. The great advantage of supercritical carbon dioxide is the relatively mild conditions required (31°C and 7.4MPa, c.f. ethanol 243°C and 6.4MPa).

ii) Freeze-drying (freezing the liquid in the gel and removal by vacuum sublimation) also avoids the capillary stress effects associated with the direct removal of liquids from gels. However, this cannot be used to produce crack-free dry monoliths, since the water crystallises and stresses the surrounding matrix in the process, leading to extensive fracturing and pore damage.

iii) Various Drying Control Chemical Additives (DCCAs) have been reported. These are discussed in Chapter 3. All of these additives modify other properties of the gels to some extent, and may therefore be undesirable and moreover difficult to remove.

iv) As discussed above, ageing gels strengthens them, increasing resistance to cracking. Although this is time consuming, it is cheap, produces more stable and reproducible gel properties and involves no contaminating additives.

v) Since it is the small pores which contribute most to capillary stress, synthesis of gels with large pores (greater than 50nm) would yield substantial benefits in reducing cracking. For example, mixing a commercial colloidal silica (Ludox®) with potassium silicate at high pH and gelling by addition of formamide gives a product in which silicate polymers are nucleated from solution by the silica colloid. Controlled pore sizes between 10 and 360nm were obtained by controlling the ratio of the two starting materials, and the samples with pore sizes greater than 60nm could be dried rapidly without cracking, even in dimensions of many centimetres.[15]

2.8   DENSIFICATION

2.8.1   Stages of Densification

Although there are many applications of silica gels prepared and dried at or near room temperature (particularly applications which involve entrapping functional organic or biological molecules within gel pores in mild conditions), heat treatment is necessary for the production of dense glasses and ceramics from gels. From the discussions so far in this chapter it is clear that, by control of the hydrolysis, condensation, ageing and drying stages, materials with a wide range of pore sizes, pore wall characteristics and general microstructure can be prepared. The detailed effects of heat treatment therefore depend on the particular characteristics of the material at the end of the low-temperature drying process. A good experimental starting point is the measurement of linear shrinkage and weight loss as a function of temperature, for samples heated at a constant rate. Commonly (Figure 2.10) three regions are observed in these experiments:

I. At low temperatures (typically <200°C) weight loss occurs as pore surface water or alcohol is desorbed, but little further shrinkage takes place, and in some cases a net expansion is observed. In this range, the skeletal structure of the silica gel behaves as a molecular solid rather than as bulk silica. The thermal expansion coefficient is typically of the order of $5 \times 10^{-5}K^{-1}$ which is close to the typical values of $10^{-4}K^{-1}$ for organic solids but two orders of magnitude larger than that for bulk dense silica. This dominates the tendency for further shrinkage as a result of increased capillary stress due to solvent loss.

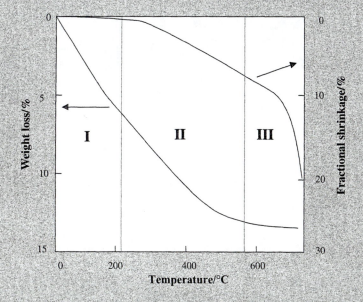

FIGURE 2.10 STAGES OF DENSIFICATION.

II. At intermediate temperatures in the range from 150–200°C to typically 500–700°C, samples generally show both weight loss and shrinkage. Three processes occur in this range: loss of organics (leading to weight loss but little shrinkage), further condensation (producing both weight loss and shrinkage) and structural relaxation (giving shrinkage with no associated weight loss). The loss of organics proves that the matrix is still porous in this stage. The spaces previously occupied by the organic species now become pores with similar size and shape to the organics themselves. These "molecular footprints" have been used for catalysis (see also chapters 6 and 7), and for generating controlled small pore sizes (e.g. by use of different alkylammonium salts). However, structural relaxation and further trans-pore condensations gradually lead to the footprints fading, particularly at elevated temperatures such as are used for catalytic processes.

III. At temperatures above the upper limit for region II behaviour, a sharp increase in shrinkage rate is observed with little or no further weight loss. The transition temperature is close to the glass transition temperature for the material, above which viscous flow occurs leading to rapid densification as thermal energy permits extended structural reorganisation. The densification process is strongly favoured thermodynamically because of the very large reduction in surface area of the material and the associated large reduction in interfacial energy. The extent of this final shrinkage, as well as the temperature at which it begins, is dependent on the structure of the material at the end of region II and thus on the conditions for all the stages in the process which have been described up to this point in this chapter. Clearly many variables are involved, and in principle this provides great scope for fine control of the onset and nature of the densification process, which has been widely exploited in ceramic processing. However, in addition to controlling all the stages which precede densification, further control of densification is possible by incorporation of viscosity modifiers whose action commences in region III. Thus, for example, increasing the alkali content of silica gels decreases viscosity in region III as the alkali exerts a depolymerising action:

$$X_3Si\text{-}O\text{-}SiX_3 + Na_2O \rightarrow 2\ X_3Si\text{-}O^-\ Na^+$$

This lowers the temperature at which densification takes place.

2.8.2 Effects of Heating Rate and Gas Evolution

It is difficult to apply conventional theories of viscous sintering to sol-gel materials because of the structural changes which occur during heating due to condensation and relaxation processes. These changes increase the viscosity if a sample is held at a steady temperature, leading to a steady decrease in the densification rate with time. Sintering rate is thus determined by the thermal and chemical history of a sample and not simply by the sintering temperature. Frequently therefore a faster densification is achieved if a sample is heated rapidly rather than sintered at a fixed temperature, the increased thermal energy compensating for the greater viscosity due to cross-linking and structural relaxation. However, there is a limit to this effect since condensation, as well as pyrolysis of any entrapped organic species, generates vapourised species at high temperature and, as densification increases, these vapours may become trapped in the solid. If vapour evolution has ceased when the pores close, further shrinkage due to continued viscous sintering will occur only until the increase in pressure due to decreased pore volume exactly balances the pressure across the curved interface as predicted from the Kelvin equation. Since the Kelvin equation predicts that ln(pressure difference) is inversely proportional to radius of curvature, whereas for an ideal gas the pressure is inversely proportional to volume, (i.e. to $1/r^3$), it is clear that the gas pressure will quite rapidly balance the interfacial pressure as pore size decreases (c.f. chapter 1, where from the Kelvin equation p_r/p_o changes from

1.01 to 3 as r changes from 100nm to 1nm, whereas the pressure increase associated with a radius decrease of a factor of 100 would be 10^6). This effect can be avoided by sintering in a vacuum or in an atmosphere of a gas which is soluble in the silica matrix. However, if vapour evolution continues after pore closure due to surface chemical reactions, such strategies will prove ineffective. The densifying gel will then cease contraction and either expand or crack to release the growing internal pressure, depending on its rigidity at that stage. For example, even completely dry aerogels can show such behaviour if the pores are small and close easily before all the remaining surface hydroxy or alkoxy groups have been removed by condensation with associated liberation of water or alcohol. Thus, acid catalysed TMOS aerogels densify but then expand whereas similar base-catalysed aerogels, with larger particle sizes and larger pores, densify more completely with no such expansion.

2.9 CONCLUSION

From this chapter it will be clear that the detailed description of the processes leading from a simple alkoxysilane precursor to a densified glass is a series of immensely complicated steps, each of which depends in some ways on what has occurred in previous steps so that the steps themselves are of increasing complexity. The fascination of the subject lies in the immense flexibility and range of possibilities arising from this complexity; its frustration is the virtual impossibility of ever ensuring that a given material is truly optimised; its solace is the knowledge that despite this, as illustrated in chapter 1, so many useful materials can nevertheless be produced in this way.

REFERENCES

1. B.K. Coltrain, S.M. Melpolder and J.M. Salva, *Proc. IV Int. Conf. Ultrastructure Processing of Ceramics, Glasses and Composites, 1989*, edited by D.R. Uhlmann and D.R. Ulrich. Wiley: New York.
2. R. Aelion, A. Loebel and F. Eirich, *J. Am. Chem. Soc.*, **72**, 5705 (1950).
3. L.C. Klein, *Ann. Rev. Mater. Sci.*, **15**, 227 (1985).
4. R.A. Assink and B.D. Kay, *J. Non-Cryst. Solids*, **99**, 359; **107**, 35 (1988).
5. A. Vioux and D. Leclercq, *Heterogen. Chem. Rev.*, **3**, 65 (1996).
6. R.J.P. Corriu, D. Leclercq, P. Lefèvre, P.H. Mutin and A. Vioux, *J. Mater. Chem.*, **2**, 673 (1992).
7. J.N. Hay and H.M. Raval, *J. Sol-Gel Sci. Technol.*, **13**, 109 (1998).
8. L. Bourget, R.J.P. Corriu, D. Leclercq, P.H. Mutin and A. Vioux, *J. Non-Cryst. Solids*, **242**, 81 (1998).
9. B. Gauthier-Manuel, E. Guyon, S. Roux, S. Gits and F. LeFaucheux, *J. Phys.*, **48**, 869 (1987).
10. P.J. Flory, *Principles of Polymer Chemistry*. Cornell U.P.: Ithaca, NY, 1953.
11. R. Zallen, *The Physics of Amorphous Solids*. Wiley: New York, 1983.
12. M. Smoluchowski, *Phys. Z.*, **17**, 557, 585 (1916); *Z. Phys. Chem.*, **92**, 129 (1917).
13. See for example: P. Meakin, T. Vicsek and F. Family, *Phys. Rev. B*, **31**, 564 (1985).
14. A.J. Vega and G.W. Scherer, *J. Non-Cryst. Solids*, **111**, 153 (1989).
15. R.D. Shoup, in *Ultrastructure Processing of Advanced Ceramics*, edited by J.D. Mackenzie and D.R. Ulrich, p. 347. Wiley: New York, 1988.

CHAPTER 3

THE CHEMISTRY OF SOL-GEL SILICATES

3.1 INTRODUCTION

As was explained in the preceding chapter the sol-gel process involves a complex pattern of chemical reactions that proceed simultaneously. This allows chemical modification in all stages of the process for the optimisation of the structure of the attained silicates to fit a particular application. These modifications can have their effect at the molecular, but also on the mesoscopic and macroscopic level and may include variation of the silicate precursor and the reaction conditions, as well as post treatment of the resulting gels and glasses.

3.2 CHEMICAL CONTROL OF THE SOL-GEL PROCESS

The properties of sol-gel materials, e.g. transparency, porosity, pore size distribution, surface functionality, strongly depend on the preparation method. By controlling the conditions during the subsequent stages of the process: hydrolysis, condensation, ageing, drying, one can fine tune the characteristics of the resulting material.

3.2.1 The Hydrolysis and Condensation Reaction

3.2.1.1 The role of the precursor molecule

Hydrolysis of the silicon alkoxide precursor is very sensitive to steric hindrance. Chain elongation and in particular chain branching in the alkoxide ligands lead to a dramatic decrease of the reaction rate.[1] The use of a higher alcohol (e.g. n-propanol) as a cosolvent leads to replacement of ethoxide ligands during the first hydrolysis step of TEOS. Consequently a significant reduction in the rate of the second reaction step is observed due to the lower reactivity of the n-propoxide groups. In general, stable modifying ligands reduce the hydrolysis and condensation rate so that less-condensed gels are obtained.[2] The replacement of the ethoxy groups in TEOS by transesterification with the bidentate 2-methylpentane 2,4-diol gives gels which have large pores and high surface areas.[3] Due to steric hindrance this precursor exhibits a low reactivity towards hydrolysis and, therefore, the structure of the gels only depends on pH and reaction temperature, rather than on the amount of water present in the reaction mixture.

A different way of modifying the gel structure through modification of the reactivity of precursors is the replacement of the alkoxy groups by other ligands such as acetate or acetyl acetone.[4] In the sol-gel processing of non-silicates such as transition metal alkoxides, which in general are more reactive towards hydrolysis, the stability of such bidentate ligands is utilised in order to retard the hydrolysis and condensation reactions. Remarkably, both the hydrolysis of TEOS in the presence of acetic acid and the hydrolysis of silicon tetraacetate lead to very fast gelation (possibly due to the reaction with ethanol to produce ethyl acetate).

Sol-gel derived materials can also be obtained from colloidal solutions, in particular colloidal silica and water glass, an approach which may be regarded as the application of precondensed starting materials.[5,6] This leads to the formation of gels with bigger pores and consequently to a reduction of stress due to the evaporation of solvent. The generation of bigger pores facilitates the drying process and hence the fabrication of thick films and bulk pieces (see also chapter 6) and this method also avoids the use of hazardous and expensive chemicals.[7]

3.2.1.2 The role of the catalyst

The hydrolysis of silicon alkoxides is dramatically promoted by the use of a catalyst. Under acid catalysis the initial step of the hydrolysis *i.e.* the conversion of the precursor molecules into the trialkoxy silanol $(RO)_3Si-OH$ proceeds rapidly. Since one of the electron donating alkoxy groups has been removed, protonation of this silanol species will be less favourable (as discussed in chapter 2) and hence the second hydrolysis step will be slower. This means that condensation reactions between Si-OH and protonated Si-OR groups of non-hydrolysed or only partially hydrolysed monomers will play a significant role in the reaction mixture. Since terminal Si-OR groups will be more reactive on both steric and inductive grounds, acid catalysed hydrolysis will initially lead to chain elongation and thereby to the formation of linear polymers. Cross-linking will predominantly occur by accidental interlinking of chains after hydrolysis of the Si-OR side-groups. Entanglement of these partially cross-linked chains leads to homogeneous, relatively dense gels with small pores.

Base catalysis, on the other hand, proceeds faster when electron donating -OR groups are removed. The consequent generation of (almost) completely hydrolysed monomers leads to cross-linking already at early stages in the process where unhydrolysed monomer is still present. Due to the high condensation rate and the interlinking of the highly cross-linked polymers, as discussed in chapter 2, a porous network is formed and gelation occurs fast.

The structure of the gel in some cases is also influenced by the particular type of catalyst. When HF is used as the catalyst instead of HCl much faster gelation times are observed. Fluoride ions can activate silicon through the formation of a stabilised hypervalent five (or even six) coordinated transition state.[8] The mechanism of catalysis is similar to that for OH⁻ ions and also leads to fast gelation times[9] (n-Bu₄NF> NaF > NH₄F > CsF > N,N-dimethylaminopyridine > N-methyl imidazole > NH₄Cl > no catalyst > HCl) and large pores.[10] Due to the fact that the Si-OH and Si-OR groups on the silica surface are largely replaced by Si-F sites, these gels are more hydrophobic, which is reflected in the reduced water uptake of the gels after drying.

3.2.1.3 The role of water

When silicon alkoxides are reacted with low concentrations of water ($H_2O/Si < 2$) initially partially hydrolysed monomers are formed which condense to form almost completely esterified — either linear or branched — polymeric species. Increasing the water content to 4–10 H_2O/Si leads to almost completely hydroxylated polymeric strands in acid catalysed systems. In base catalysed gels this leads to the generation of a fully hydrolysed polymer in the presence of still unhydrolysed monomer which induces phase separation in many cases. Consequently aqueous regions solvating the hydrophilic polymer and solvent rich regions with unreacted hydrophobic monomer are formed.[11] A further increase of the water concentration (25–50 H_2O/Si) leads to separation of the individual polymeric strands and hinders the intermolecular condensation reactions. This leads to the formation of isolated cyclic structures and consequently to the formation of more dense spherical particles.[12] In general, the presence of surface silanol groups promotes gelation by increasing the condensation rate.[13]

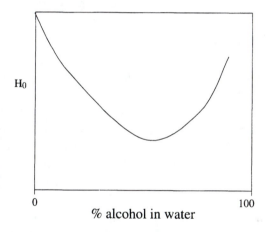

FIGURE 3.1 A GENERALISED REPRESENTATION OF THE RELATION BETWEEN THE ACTIVITY H_0 AND THE ALCOHOL CONTENT IN AQUEOUS SYSTEMS.

Addition of solvents such as alcohols, THF and dioxane affects the structure of water by breaking the hydrogen bonds. This increases the solvation of H^+ and OH^- ions and thereby decreases, respectively, the acidity or basicity of the solution and consequently the rate of hydrolysis. However, since the relation between the water/alcohol ratio and the activity of the catalyst follows a "U"-curve (figure 3.1), for systems containing more than 60% alcohol the reverse effect is observed. In these cases increasing the water concentration increases the activity of the catalyst and hydrolysis is promoted.[14]

The addition of hydrogenated sugars (e.g. mannitol, sorbitol) decreases the activity coefficient of water by improving the hydrogen bonding structure. This increases the acidity (or basicity) of the solution and consequently the rate of hydrolysis of TMOS, in the case of HCl catalysis the gel time was reduced from 720 hr to 15 min.[15]

3.2.2   Gelation

At neutral pH the polymerisation reaction proceeds through nucleophilic attack of the Si-O⁻ group. The nucleophilicity of this group can be fine-tuned through addition of solvents. Polar protic solvents such as formamide and alcohols lower the reactivity through the formation of hydrogen bonds. This decreases the efficiency of the condensation process leading to more highly branched structures and consequently to reduced gel times. The replacement of an alcoholic solvent by dioxane leads to activation of the Si-O⁻ group. Due to the efficiency of the condensation process dense gels are obtained. In general increasing amounts of formamide lead to a coarsening of the gel structure and to an increase of the pore size.

By acting as a hydrogen bond acceptor formamide and dimethylformamide decrease the acidity of the reaction mixture and thereby the hydrolysis rate in acid catalysed reactions. For the same reasons the nucleophilicity of the silanol groups is enhanced, leading to faster condensation.[16]

It must be noted however that in an acid catalysed system formamide is hydrolysed to yield formic acid and ammonia according to the following reaction:

$$H\text{-}(CO)\text{-}NH_2 + H_3O^+ \rightarrow H\text{-}(CO)\text{-}OH + NH_4^+$$

consequently leading to an increase of the pH during the sol-gel process.[17] This gives the process initially the character of an acid catalysed process [fast (partial) hydrolysis of *all* the precursor molecules] and later of a base catalysed process, leading to high interconnectivity of small oligomers, and hence to a decrease of the gelation time.[18]

It was found for DMF and acetonitrile that by hydrogen bonding to the silanol groups on the surface these polar aprotic solvents inhibit the coalescence of the silica particles.[19] The same effect was found for the addition of ethylene glycol and ethylene glycol monomethylether which can bind to the surface both by hydrogen bonding and by the formation of silyl ester bonds.[20] This results in the formation of larger particles and hence in a larger pore size. It was shown for formamide that the addition of solvents also changes the polarity of the reaction mixture and therefore affects the time needed for a phase separation to occur.[21]

3.2.3   Ageing

During the ageing process the gel hardens by the formation of cross-links. However, the number of new Si-O-Si bonds responsible for this rigidification was found to be remarkably low. The residual hydroxyl content of the investigated silica gels amounted to ca 0.5 OH/Si.[22]

The pore structure of sol-gel silicates can be modified by aging them in solvents[23] or solutions other than the mother liquor. The rate of ageing is directly related to solubility: higher temperatures, hydroxide and fluoride ions increase the rate of aging (and thereby the development of the pore size and the pore size distribution).[24] By dissolution and reprecipitation, neck formation occurs between particles, which leads to strengthening of the pore wall preventing collapse of the pores during the drying process (see Chapter 2).

Ageing in ethanol leads to (partial) esterification of the silica surface — increasing the number of Si-O-R groups — as well as to depolymerisation of the silica network, *i.e.* by the breaking of Si-O-Si linkages. Both reactions lead to an increase of the surface area, the former by increasing the surface roughness through the introduction of the bulkier ethoxy groups, the latter by creating new surfaces. Immersion in water of a previously unwashed gel leads to the phase separation of monomeric and oligomeric silica species, causing the gels to turn opaque. Washing with ethanol can be used to remove these low molecular weight reaction products after which the gels remain optically transparent upon ageing in water. This leads to a reduction of the surface area by the complete hydrolysis of surface Si-O-R groups and the consequent formation of new Si-O-Si bonds and yields materials with smaller pore sizes and a broader pore size distribution. Both processes are reversible and are partially preserved in the dried gels. Successive alternations of the pore fluid only leaves the effect of the final one.

3.3 ADDITIVES FOR STRUCTURING AND PROCESSING

3.3.1 Drying Control Additives

One of the most serious problems encountered when one attempts to produce gel monoliths through the sol-gel process using silicon alkoxides as starting materials is fracture and crack formation during the drying of these gels. This fracturing is caused by stress due to changes in pore size during the process and by capillary forces arising from evaporation of the solvent. In order to overcome these problems a number of methods have been developed. The first is the "slow rate evaporation method" where the removal of the solvent is slow and carefully controlled.[25] This gives good results but requires very long processing times. The second method is known as supercritical drying, in which the solvent is removed under controlled pressure.[26] This is the fastest and most reliable process to date but has the disadvantage that it is a discontinuous process which limits its industrial application. The third process is the use of drying control chemical additives (DCCA's) such as formamide,[27] dimethylformamide (DMF),[28] acetonitrile,[29] and oxalic acid.[30] Addition of these agents allows drying at elevated temperatures and ambient pressures without crack formation.

The action of DCCA's leads to a change in the structure of the sol-gel material. As was described in the previous section this proceeds via a number of complex mechanisms in which all stages of the sol-gel process *i.e.* hydrolysis, condensation, gelation, phase separation and aging, are influenced. The mechanisms of crack prevention are equally complex.

Firstly the DCCA (*e.g.* formamide, DMF) causes the generation of large pores with narrow size distribution, as was described in the preceding section. Due to the larger pore size, capillary forces will be weaker and hence the stress exerted on the gel during drying will be smaller. Furthermore, since the vapour pressure is higher in larger pores, according to the Kelvin equation, the increased pore size promotes the evaporation of the solvent.[31]

The second mechanism of action is the binding of the DCCA to the silica surface through the formation of hydrogen bonds. This will facilitate the removal of water molecules by preventing their interaction with the silanol groups on the pore walls. Additionally, the remaining pore fluid will mainly consist of the DCCA which generally has a lower surface tension than the original aqueous phase.

Crack formation can also be avoided by the addition of surfactants to the sol-gel mixture.[32] By adhering to the particle surface the surfactant influences the pore formation, but more importantly it modifies the inner pore surface with alkyl chains. These hydrophobic groups decrease the interaction of the water with the pore wall and thereby reduce drying stress.

3.3.2   Organic Templates

When organic molecules other than solvent are added to a sol they become entrapped upon gelation and will be retained in the xerogel. This procedure can be used to add structure to the gel. The structuring of silica at the molecular level through the interaction with organic additives is beautifully exemplified by the formation of a sandwich-like composite in which two cyclodextrin molecules enclose a double six-ring silicate (See figure 3.4).[33]

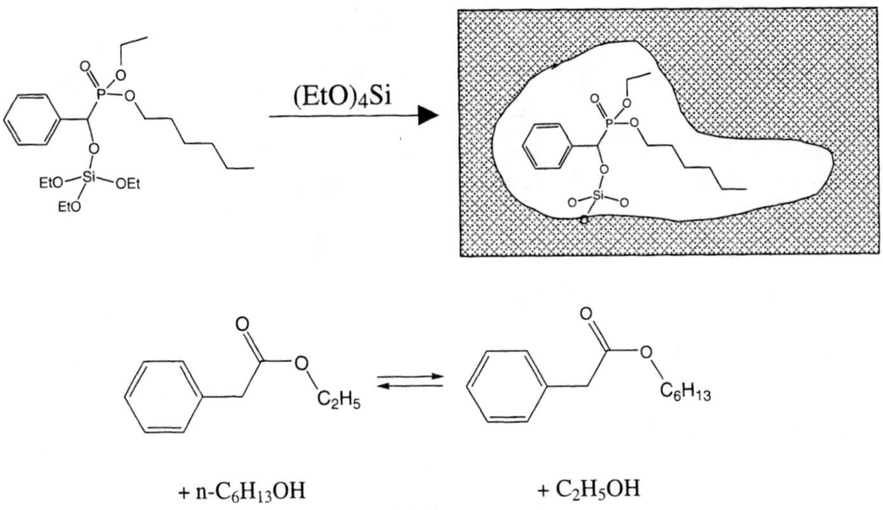

FIGURE 3.2   IMPRINTING OF A PHOSPHONATE WICH ACTS AS A TRANSITION STATE ANALOGUE FOR THE TRANS ESTERIFICATION OF ETHYL BENZOATE WITH HEXANOL.

The molecular structures of sol-gel materials can also be modified by using the shape and structure of organic molecules to mould the inorganic phase. Like other polymeric materials, sol-gel derived silicates can be imprinted using molecules that mimic the transition state of a selected organic reaction. The phosphonate (figure 3.2) which serves as an analogue of the transition state in the transesterification of ethyl benzoate, was immobilised in a sol-gel matrix. After aging and drying of the gel and subsequent removal of the template the phosphonate had left a precise imprint that selectively catalysed the transesterification with hexanol.[34]

The lyotropic liquid crystalline phases formed by surfactants can be used for the structuring of the mesopores of inorganic sol-gel derived materials. The hexagonal phase, formed upon self-assembly of cetyltrimethylammonium surfactant molecules into rod-like micelles was used as a template for the structuring of aluminosilicates (figure 3.3).[35] Due to the electrostatic interactions between the cationic surfactant and the negatively charged precursors the inorganic phase is deposited around the aggregates, forming a honeycomb-like structure. When the organic material is removed by heating to 540°C a hexagonal array of uniform channels is obtained. The diameter of the channels can be tuned either by changing the length of the alkyl chains of the templating amphiphiles, or by adding auxiliary hydrocarbons such as trimethylbenzene.

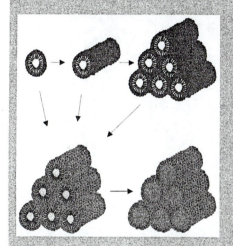
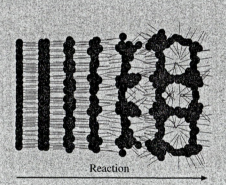

Reaction

FIGURE 3.3 THE TEMPLATING PROCESS AND FORMATION OF THE POROUS NETWORK IN MCM TYPE MATERIALS.

After the mixing of the organic and inorganic components a lamellar phase is initially formed, which gradually transforms into a hexagonal structure.[36] It is proposed that in the early stages of the synthesis the presence of highly charged silica oligomers decreases the repulsive forces between the positively charged surfactant head groups. This reduces the molecular area and permits the formation of a lamellar phase. As the polymerisation proceeds, Si-O-Si bonds will be formed reducing the amount of free Si-O$^-$ groups and transforming the surfactant aggregates into a hexagonal pattern of micellar rods.

Soon after the discovery of these so-called M41S materials the approach was generalised and mesoporous materials based not only on hexagonal, but also lamellar[37] and cubic structures[38] and derived from cationic, as well as anionic[39] and non-ionic[40] surfactants became available. The inorganic phases were extended to silicon-,[41] titanium-,[42] tantalum-,[43] manganese-,[44] niobium-,[45] and vanadium-oxide[46] and also chromium,[47] copper[48] and zinc[49] have been incorporated. Self-assembled structures of organic molecules can also be used to generate a scaffold for the structured growth of the inorganic phase on a macroscopic scale. The basic examples are the mineralisation of large organic structures such as the use of bacteria as templates for the formation of fibers[50] and the deposition of silica on a helical ribbon composed of aggregating phospholipid molecules.[51] A somewhat more complex and less well-understood example is the generation of square, tube-like silicate structures that are formed upon hydrolysis of TEOS in the presence of DL-tartaric acid.[52] It is proposed that this racemic mixture, as opposed to the pure enantiomer (or the meso compound), self-assemble to form a ladder-polymer which structures the gel leading to these remarkable structures.

a   **b**

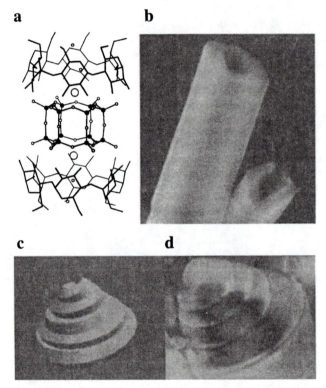

c   **d**

FIGURE 3.4 STRUCTURING OF SILICA ON A MOLECULAR AND A MACROSCOPIC LEVEL. (A) THE X-RAY CRYSTAL STRUCTURE OF THE SANDWICH-LIKE COMPOSITE CONSISTING OF 2 CYCLODEXTRIN MOLECULES AND A DOUBLE SIX RING SILICATE ($\bullet$ = SI, $\circ$ = O)[30]. (B) HOLLOW SILICATE TUBES FORMED VIA THE ADDITION OF DL-TARTARIC ACID.[49] (C) PREDICTED AND (D) EXPERIMENTALLY OBSERVED MORPHOLOGIES OBTAINED THROUGH THE USE OF LYOTROPIC LIQUID CRYSTALLINE PHASES AS TEMPLATING AGENTS FOR THE SOL-GEL PROCESS.[50]

But the most impressive examples of macroscopic structuring were obtained by using different aggregates of cetyltrimethylammonium chloride to template and sculpture the formation of silicate structures (figure 3.4). By growth-driven curvature processes, a large number of morphologies, such as gyroids, knots and spheroids, were obtained which resemble biological shapes.[53]

3.4   ORMOSILS

3.4.1   Modified Precursors

The introduction of a non-hydrolysable substituent on the silicon atom of a precursor has a dramatic effect on the material obtained after condensation of the hydrolysed monomers. Modified silicate precursors with various organic substituents, both functional and non-functional, have been used to prepare a whole range of materials with different structures and properties. This has been done by the polymerisation of the pure (hydrolysed) monomers, as well as by the co-condensation of a (generally) trifunctional and a tetrafunctional precursor such as TMOS or TEOS.

Although hydrolysis and condensation of methyltriethoxysilane is fast under acidic conditions — due to the lower steric hindrance of an alkyl group compared to an alkoxy group[54] — gelation is retarded compared to reaction mixtures containing only TEOS. This is caused not only by the inherent lower degree of connectivity upon condensation; polymerisation of hydrolysed methyltriethoxysilane only proceeds via the elimination of water, not by alcohol formation.[55] Consequently it was found that for mixtures of TEOS and methyltriethoxy silane the formation of longer polymeric strands was inhibited by the presence of alcohol, which favours the formation and precipitation of crystalline silsesquioxane (T_8) cages. This was hampering the preparation of optically transparent materials from organotrialkoxy silanes via room temperature sol-gel processing. The problem was overcome by the application of a two-stage hydrolysis and condensation procedure.[56] By partial hydrolysis of the precursor molecules and subsequent evaporation of the ethanol the formation of crystalline cage compounds was prevented. Further hydrolysis of the reaction mixture led to the formation of transparent films.

The structures with the formula $R-SiO_{3/2}$ obtained from the hydrolysis and condensation of organotrialkoxysilanes are generally referred to as silsesquioxanes.[57] The bulk of the investigations into the preparation and structures of these materials has focused on polyphenyl- and polymethyl silsesquioxanes. It is proposed that the condensation reaction proceeds via the formation of more or less well-defined cage structures that under strong hydrolytic conditions convert into polymers with molecular weights of 10^3-10^5 (figure 3.5). These polysilsesquioxanes possess a ladder structure that only shows significant branching at higher ($\sim10^5$) molecular weights. Sol-gel processing of materials containing aliphatic substituents bigger than a methyl group however leads to the formation of incompletely condensed branched structures (T-resins), or to various cage compounds.

Bridged polysilsesquioxanes are obtained from precursors in which the organic fragment is connected by means of C-Si bonds to two or more trifunctional silyl groups (figure 3.6).[58] Condensation of the hydrolysed monomers leads to rapid growth of a heavily branched network which gels at extremely low concentrations. In this way highly porous materials are formed in which the organic linker forms an integral part of the three-dimensional network. Due to the low monomer con-

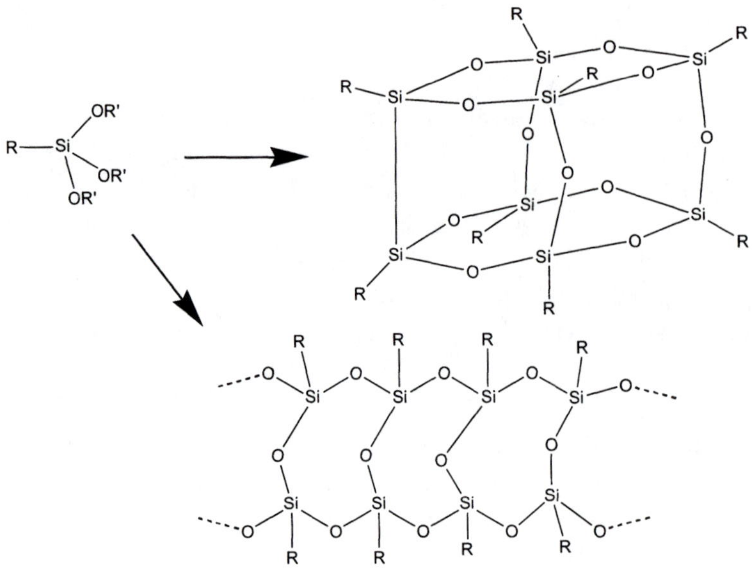

FIGURE 3.5 CAGE COMPOUNDS AND LADDER POLYMERS FROM SILSESQUIOXANES.

centrations used and the fact that 6 molecules of alcohol are released upon the hydrolysis of every precursor molecule, these gels undergo shrinkage of 90–95% upon drying.

Nevertheless, when DCCA's are used crack-free monolithic materials can be obtained. By variation of the linkers these gels can be prepared with a wide range of ratios of organic and inorganic material without phase separation. The porosity of the materials depends both on the size and rigidity of the organic component. Highly porous gels can be obtained from precursors with short alkyl linkers; chain elongation leads to a decrease in porosity due to collapsing of the more flexible aliphatic chains. However, when rigid alkynylene spacers are used, highly porous materials are obtained.

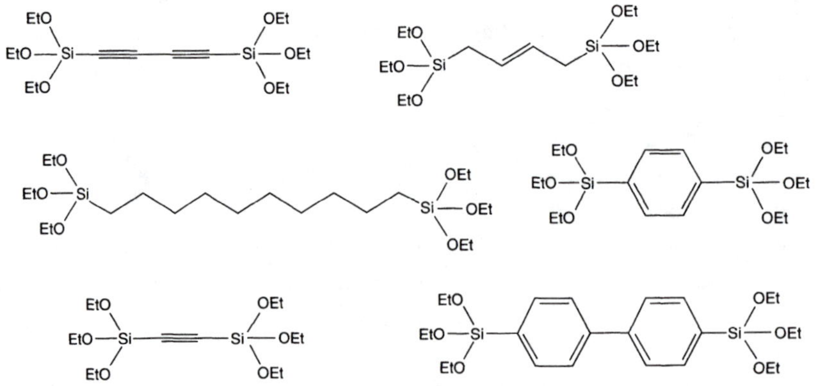

FIGURE 3.6 EXAMPLES OF BRIDGED SILSESQUIOXANES.

In order to fine-tune their properties, sol-gel materials have been prepared by the hydrolysis and co-condensation of mixtures of tetrafunctional precursors (e.g. TMOS, TEOS) and various organotrialkoxysilanes. It is obvious that the introduction of a non-hydrolysable substituent will lower the connectivity of the network by leaving "dead ends". The presence of these "dead ends" will affect the chemical composition of the pore surface, which means that the choice of the organic substituents used will determine the character of the resulting material. Organotrialkoxy silanes have been used to obtain sol-gel derived materials with additional functionality. The sol-gel processing of polymerisable, fluorinated or metal binding organotrialkoxy silanes leads to the formation of composite materials which will be discussed in the following sections.

3.4.2 Entrapment of Functional Materials

3.4.2.1 Efficiency of entrapment

The fact that the sol-gel route is a low temperature method allows doping of inorganic materials with organic compounds, thereby adding functionality to the resulting glass. Organic molecules added to the sol become entangled in the developing network and consequently entrapped in the final xerogel (figure 3.7). However, when these gels are used in liquids it is important to note that the efficiency of the

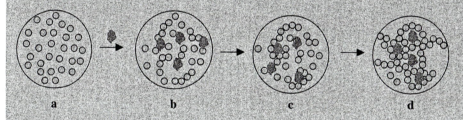

FIGURE 3.7 THE ENTRAPMENT OF GUESTS IN A SILICA MATRIX. A) FORMATION OF SOL PARTICLES DURING THE HYDOLYSIS AND CONDENSATION STAGES. B) ADDITION OF THE GUEST TO THE SOL. C) TRAPPING OF THE GUEST BY THE GROWING SILICA NETWORK. D) IMMOBILISATION OF THE GUEST IN THE SILICA MATRIX.

entrapment depends on a number of factors, such as the pore size distribution and porosity of the gel, the size of the dopant, and the development and stability of the matrix network. In order to prevent leaching into the liquid phase, dopants have been modified with trialkoxysilane groups which allows incorporation in the developing polymer network during the sol-gel process. The two main methods for derivatisation are (1) the reaction with an alkyltrialkoxy silane bearing a suitable functional group, e.g. an amino, epoxy or vinyl group, or (2) hydrosilylation of a double bond in the dopant molecule with a trialkoxy silane (figure 3.8).[59]

The special features of sol-gel materials i.e. the fact that they are porous and generally optically transparent, have opened the way for many possible applications of these composites which will be discussed in more detail in Chapter 6.

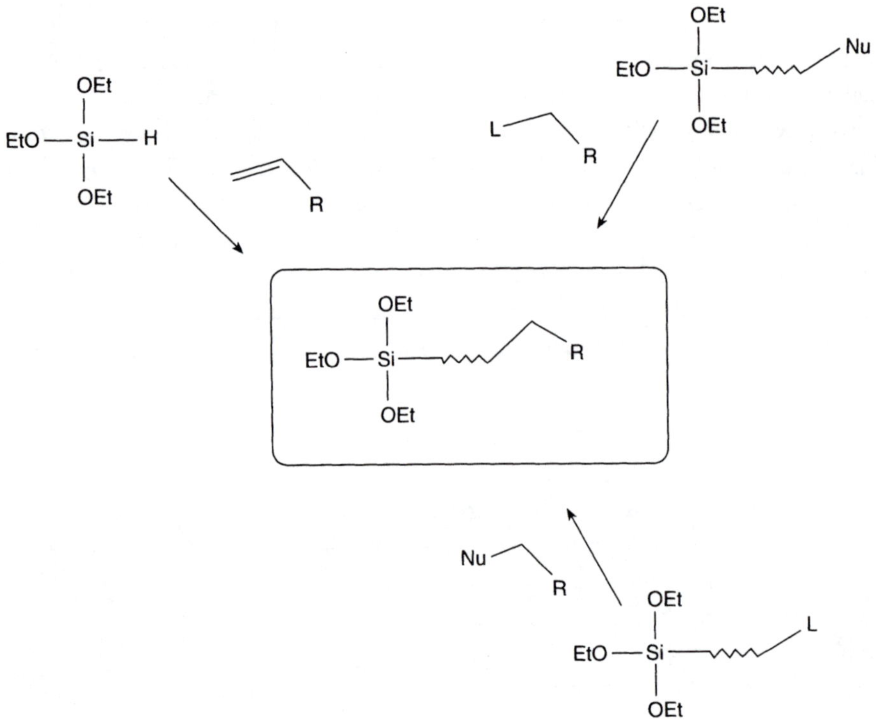

FIGURE 3.8 THE SYNTHESIS OF TRIETHOXYSILANE DERIVATIVES FOR THE COVALENT IMMOBILISATION OF FUNCTIONAL GROUPS IN A SOL-GEL MATRIX.

3.4.2.2 Photochemistry and photophysics inside sol-gel matrices

The fact that xerogels with entrapped organic molecules can be obtained as transparent materials, makes them extremely valuable for studies directed towards the storage of light energy in the form of charge-separated ion pairs. The charge separation between the photogenerated pyrene cation (Py^+) and the methyl viologen cation (MV^+) normally is rapidly quenched by the back reaction. Avnir et al. have entrapped pyrene and methyl viologen in a sol-gel matrix.[60] A mobile charge carrier (N,N'-tetramethylene 2,2'-bipyridinium bromide; TMP^{2+}) dissolved in the aqueous phase (which occupied the pores) was used for the chemical communication between

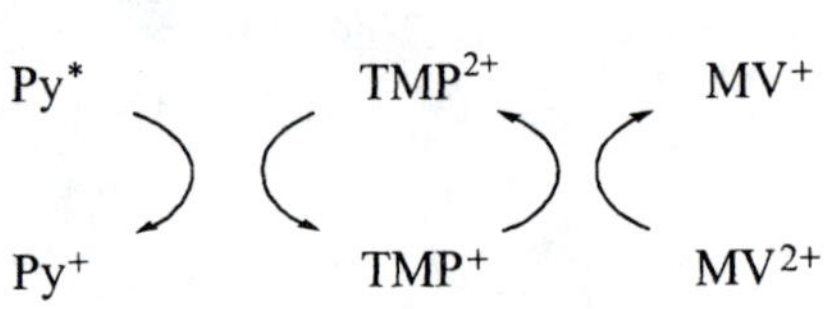

FIGURE 3.9 USE OF A MOBILE REDOX MEDIATOR TMP^{2+} TO EFFECT PHOTO-EXCITED CHARGE TRANSFER BETWEEN SEPARATED PYRENE AND METHYL VIOLOGEN MOLECULES IN A SOL-GEL MATRIX[60].

the two entrapped species. Upon irradiation, a Py^+/MV^+ cation pair was formed which had a lifetime of several hours. In a similar manner a Ir(II)/Ru(II) system with dimethoxy benzene as the shuttle was used to generate hydrogen gas by reduction of H_3O^+.[61]

The preparation of photochromic glasses, i.e. glasses that change colour upon exposure to light, by traditional methods has always been limited to the use of the few dopants that could withstand the high temperatures involved in the melting process. By utilising the sol-gel process a vast number of organic photochromic dyes have become available for this application (see chapter 6 for more details). An alternative approach to limiting optical exposures involves the use of reverse saturable absorbers. A particularly nice example is the use of trialkoxysilane derivatised Buckyballs (C_{60}) for optical limiting glasses for laser flash protection.[62] These molecules are easily excited to a singlet state which rapidly converts to a long-lived triplet state. The C_{60} derivatives show a triplet-triplet transition between 700 and 750 nm with a higher absorption coefficient than that of the initial singlet excitation. Hence at high light intensities, giving a high population of the lower triplet state, the total absorption coefficient for these wavelengths becomes very high, providing flash protection.

3.4.2.3 Bioactive glasses

In the search for functional sol-gel materials the most impressive achievement has been the encapsulation of bioactive materials. The entrapment of yeast cells yielded a material with invertase activity, but encapsulated cell-free invertase did not show any activity, most probably due to the denaturating conditions used in the process.[63] The next step was the immobilisation of alkaline phosphatase, which did retain its activity but precipitated during the process.[64] However, by the replacement of the low molecular alcoholic solvents for less denaturating polyethylene glycols, Braun et al.[65] succeeded in the encapsulation of various enzymes such as trypsin, aspartase and peroxydase without precipitation and with retention of activity. The addition of polyethylene glycol causes the sol-gel mixture to separate into a denaturing organic phase consisting of TMOS saturated with PEG and traces of water, and an aqueous phase containing hydrolysed silicate polymers, PEG and the enzyme. After gelation of the aqueous phase the organic phase is removed and the remaining gel is incubated at 4–7°C and eventually dried at 30–37°C. A further improvement was the two stage method developed by the group of Zink.[66] They carried out an acid hydrolysis of TMOS, without the addition of a solvent, by sonication of the reaction mixture. The obtained homogeneous sol was buffered at pH 6.0 before the enzyme solution of the same pH was added. In this way transparent homogeneous gels with catalytic activity were obtained.

3.4.2.4 Clustering and isolation of inorganic dopants

Sol-gel materials have also been used as matrices for the growth of inorganic particles of defined size. Examples include the formation of microcrystals of semiconductor materials, noble metals and catalysts. This method provides materials in which the particles are dispersed homogeneously and frequently have diameters < 10 nm which allows for quantum effects. For example gold particles of controlled size can be generated through photodeposition of a soluble precursor [dimethyl (hexafluoro acetylacetonato)-gold or dimethyl(trifluoroacetyl-acetonato)gold] inside optically transparent sol-gel silicates.[67] The size of the particles was found to be dependent on the irradiation time since irradiation promotes their aggregation. Particles of a desired size can be obtained by monitoring the absorption spectra during the irradiation. In contrast to colloidal solutions in which gold particles generally are only stable for a few weeks unless the preparation is done with extreme care, the obtained materials were stable over a period of two years.

The doping of sol-gel glasses with isolated organometallic compounds has been performed in two ways. The first way is through the addition of the complex to the sol, leading to entrapment during the network formation. The second is through the co-condensation of a trialkoxysilane-derived ligand and a tetrafunctional precursor. In this manner the complex is formed via ligand exchange after the gel has formed. The properties of the above mentioned materials and their application in the field of catalysis and chemical sensing will be discussed in chapter 6.

3.5 HYBRID MATERIALS

In order to provide polymeric materials with the physical and mechanical properties required for applications, most of them are used as blends or composites. Since the properties of composite materials depend not only on the properties of the individual components, but also on the phase morphology and particularly on the interfacial properties, the degree of phase separation is of paramount importance. The sol-gel method allows for nanoscale mixing of organic and inorganic polymers, in some cases even on a molecular level. Because of this intimate mixing, these so-called hybrid materials often are highly transparent as compared to other composites, since light scattering due to domain formation is strongly reduced. Five major classes of hybrids have been described: (1) Pre-formed soluble organic polymers entrapped in a xerogel; (2) preformed embedded polymers with covalent links to the inorganic network; (3) mutually interpenetrating organic and inorganic networks; (4) interpenetrating networks with covalent links between the organic and inorganic phases; (5) non-shrinking sol-gel composites (figure 3.10).[68]

The main problem with the preparation of hybrid materials from preformed polymers is the solubility of these macromolecules in solvent (mixtures) that are compatible with the sol-gel process. This is complicated by the liberation of alcohol during the hydrolysis of the precursor molecules which may cause precipitation of the polymer during the reaction. Transparent homogeneous hybrids of type (1) have been obtained for a number of polymers with basic functional groups, such as polyamines and poly(vinyl pyridines), which remain soluble in hydrolysing acidic sols.

Type (2) materials are generally obtained from the hydrolysis of a tetraalkoxysilane in the presence of a soluble preformed polymer functionalised with trialkoxysilane groups. These again can be prepared by hydrosilylation of terminal double bonds or by coupling reactive groups in the polymer with functionalised trialkoxysilanes.

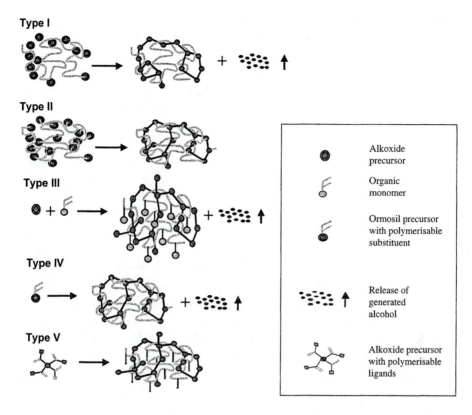

Type I

Type II

Type III

Type IV

Type V

⊙	Alkoxide precursor
	Organic monomer
	Ormosil precursor with polymerisable substituent
	Release of generated alcohol
	Alkoxide precursor with polymerisable ligands

FIGURE 3.10 THE 5 CLASSES OF ORGANIC-INORGANIC POLYMERIC HYBRID MATERIALS ACCORDING TO NOVAC.[58]

Other ways of preparing these materials are by co-condensation of a tetraalkoxysilane and a hydroxy- or alkoxy-silane terminated polymer, (such as polydimethyl siloxane and aminopropyltriethoxysilane-endcapped poly(methyl oxazoline), respectively), or by the use of hydroxylated polymers such as polyvinyl alcohol and cellulose, which become linked to the network during the drying process through the formation of silylether bonds.

Since only a limited number of polymers are soluble in the tricomponent sol-gel solution much effort has focused on the simultaneous formation of an organic and an inorganic matrix. Aiming at new materials for contact lenses, Schmidt synthesised high quality composites with excellent hardness, transparency and wettability from the co-condensation of 3-glycidoxypropyltrimethylsilane and titanium alkoxides.[69] However, due to the ionic character of the glass it was very brittle. This problem was solved by increasing the organic content of the material through the addition of methacrylates together with a trialkylsilyl-functionalised propylmethacrylate to the polymerising mixture, resulting in a type (4) material with the necessary elasticity for manufacturing and handling. In Chapter 6 the choice of components for the fabrication of such contact lenses will be discussed in more detail.

Although the formation of simultaneous interpenetrating networks solved the problem of solubility and provided homogeneous interpenetrating phase morphology, the shrinking of the sample during drying still remained a problem. This problem is caused by the release of the alcoholic ligands during the hydrolysis process. Non-shrinking hybrids (type (5) materials) could be obtained by replacing these ligands by polymerisable alcohols. Upon hydrolysis of this precursor two polymerisable components, an inorganic and an organic monomer, are generated simultaneously and no evaporation is necessary. By choosing the appropriate ligands for silicon the organic/inorganic ratio can be fine-tuned. Non-shrinking hybrids with glass contents up to 50% were obtained using poly(silylic acid) oligomers esterified with a polymerisable alcohol.

3.6 SURFACE MODIFICATION

After ageing of a tetraalkoxysilane derived gel and the subsequent drying to either a xerogel or an aerogel, a porous material remains consisting of interlinked SiO_4 tetrahedra. At the pore surface the structure can terminate in either Si-O-Si groups with the oxygen link pointing to the surface, or in several types of Si-OH groups. Depending on the substitution pattern of silicon the silanol groups may be classified as geminal (two hydroxy groups on the same silicon atom), vicinal (on neighbouring silicon atoms with one bridging oxygen atom) and isolated (on a silicon atom not linked to any silanol groups).

The polarity and reactivity of these surfaces largely depends on the ability of these groups to interact with each other, i.e. by the formation of hydrogen bonds. Hydrogen bond formation depends predominantly on the distance between two silanol groups and on their mobility. Geminal hydroxy groups are too close to form hydrogen bonds, whereas isolated silanols generally are too far apart. Vicinal silanols can be involved in different types of hydrogen bonds. However, one should bear in mind that the silica surface mostly (but not entirely) consists of closed rings of interlinked SiO_4 tetrahedra with varying ring sizes. The mobility of the silanol groups and therefore their ability to form hydrogen bonds, is determined by the connectivity at the surface. Moreover, the curvature of the surface also plays an important role in determining the distance between two adjacent hydroxy groups: on concave surfaces the hydrogen bonding generally is stronger than on flat or convex ones.

3.6.1 Dehydroxylation

The large number of surface hydroxyl groups in porous sol-gel derived materials is, together with the high surface areas, responsible for the large quantities of surface absorbed water present in these gels. This interferes with the transparency of the materials, which is an essential requirement for their application in optical devices. The elimination of hydroxy groups at the surface in favour of the formation of Si-O-Si linkages is a common method to increase the optical quality of sol-gel derived materials. This can be achieved either thermally or chemically. Thermal dehydroxylation

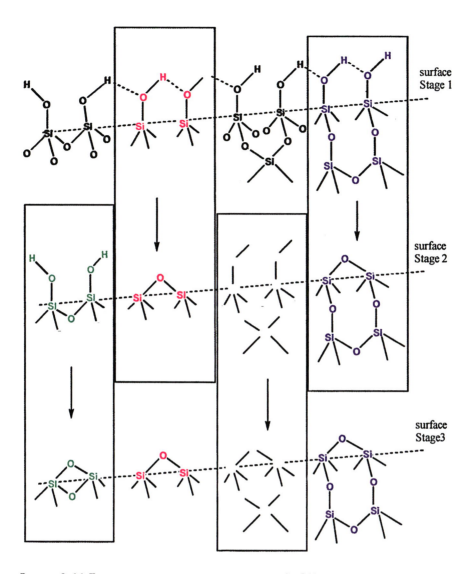

FIGURE 3.11 THE DEHYDROXYLATION OF DIFFERENT SI-OH GROUPS.

generally involves heating of the gel to temperatures up to 800°C. In this process first physisorbed water is lost and subsequently vicinal silanol groups are eliminated with the formation of rings (figure 3.11). The elimination of geminal and isolated Si-OH groups only occurs at higher temperatures since rearrangement of the silica network is required, in which they become adjacent to another silanol group.

Since thermal dehydroxylation often leads to bloating and foam formation, chemical methods are also employed to remove surface silanol groups. The most frequently used method is treatment with chlorine, which removes the hydroxyl groups according to the reaction:

$$Si\text{-}OH + Cl_2 \quad \rightarrow \quad Si\text{-}Cl + \tfrac{1}{2}O_2 + HCl$$

This treatment is normally preceded by a high temperature oxygen treatment in order to remove all remaining organics. Other chlorination reagents are CCl_4, $SiCl_4$ and $SOCl_2$. However, upon densification of the glass at high temperatures, some bloating can still occur as a consequence of incomplete dehydroxylation, leading to HCl elimination by Si-O-Si bond formation. This problem can be overcome by substituting hydroxyls by fluoride, instead of chloride. Fluoride has a much higher affinity for silicon and is therefore not eliminated by the reaction:

$$Si\text{-}OH + Si\text{-}X \rightarrow HX + Si\text{-}O\text{-}Si \qquad X = Cl, F$$

Reagents used for the fluorination of porous silicates are HF, NH_4F, and SiF_4.

3.6.2 Surface Functionalisation

Co-condensation of alkyltrialkoxy silanes and tetraalkoxysilanes is a common method for the introduction of functional groups on the pore surface (figure 3.12). The main drawback of this method is the fact that the functionalised precursor is distributed

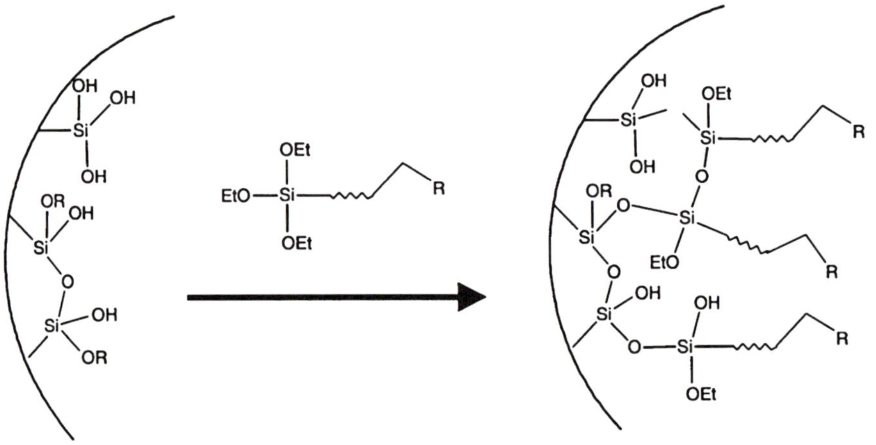

FIGURE 3.12 FUNCTIONALISATION OF A SILICA SURFACE USING AN ALKYL TRIALKOXY SILANE.

throughout the entire gel, and not only on the pore surface.[70] This leads to materials with variable surface functionalisation and low interconnectivity in the bulk. An alternative method is the modification of a gel by reacting its surface silanol groups with a precursor carrying the desired surface functionality.[71] For this purpose both alkoxy and chloro-alkylsilanes have been used.

When an alkyl silane reagent is deposited from aqueous solvents, hydrolysis and condensation lead to the formation of oligomers, which complicate the formation of a monomolecular layer.[72] For this reason generally organic solvents are used. However, alkyl trialkoxysilanes as such do not react with surface silanol groups. This problem is overcome by the in-situ reaction with water adsorbed on the silica surface. In this way the precursor is hydrolysed prior to deposition, but without interference of the undesired polymerisation. The various reagents used for surface functionalisation will be discussed in more detail in Chapter 6.

REFERENCES

1. R. Aelion, A. Loebel, F. Eirich, *J. Am. Chem. Soc.*, **72**, 5705 (1950).
2. J. Livage, M. Henry, C. Sanchez, *Prog. Solid State Chem*, **18**, 259 (1988).
3. G. Zehl, S. Bischoff, F. Mizukami, H. Isutzu, M. Bartoszek, H. Jancke, B. Lucke, K. Maeda, *J. Mater. Chem.*, **5**, 1893 (1995).
4. C. Sanchez, F. Babboneau, S. Doeuff, A. Leasutic in *Ultrastructure Processing of Ceramics Glasses and Composites*, edited by J.D. Mackenzie and D. Ulrich. San Diego (1986).
5. W. Cao, R. Gerhardt, J.B. Wachtman Jr. *J. Amer. Cerm. Soc.*, **71**, 1108 (1988).
6. M.F.M. Zwinkels, S.G. Järås, P. Govind Menon, K.I. Åsen, *J. Mater. Sci.*, **31**, 6345 (1996).
7. M. Toki, S. Miyashita, T. Takeuchi, S. Kanbe, A. Kochi, *J. Non-Cryst. Solids*, **100**, 479 (1988).
8. R. Winter, J.-B. Chan, R. Frattini, J. Jonas, *J. Non-Cryst. Solids*, **105**, 214 (1988).
9. R. Winter, J.-B. Chan, R. Frattini, J. Jonas, *J. Non-Cryst. Solids*, **105**, 107 (1988).
10. E.M. Rabinovich, *J. Non Cryst. Solids*, **100**, 174 (1988).
11. C.J. Brinker, K.D. Keefer, D.W. Schaefer, R.A. Assink, B.D. Kay, C.S. Ashley, *J. Non-Cryst. Solids*, **63**, 45 (1984).
12. I. Strawbridge, A.F. Craievich, P.F. James, *J. Non-Cryst. Solids*, **72**, 139 (1985).
13. A.H. Boonstra, T.N.M. Bernards, *J. Non-Cryst. Solids*, **108**, 249 (1989).
14. R.P.J. Corriu, D. LeCleck, A. Vioux, M. Pauthe, J. Phalippou, in *Ultrastructure processing of advanced ceramics*, edited by J.D. Mackenzie and D. Ulrich, p. 113–126. Wiley: N.Y. (1986).
15. J.P.H. Boyer, R.P.J. Corriu, R.J.M.Perz, C.G. Teye, *Tetrahedron*, **31**, 2075 (1975).
16. G. Orcel, L. Hench, *J. Non-Cryst. Solids*, **79**, 177 (1986). G. Orcel, J. Phalippou, L. Hench, *J. Non-Cryst. Solids*, **104**, 170 (1988).
17. H. Kaji, K. Nakanishi, N. Soga, *J. Non-Cryst. Solids*, **181**, 16 (1995).
18. N. Viart, J.L. Rehspringer, *J. Non-Cryst. Solids*, **195**, 223 (1996).
19. I. Artaki, T.W. Zerda, J. Jonas, *J. Non-Cryst. Solids*, **81**, 381 (1986).
20. T. Katagiri, T. Maekawa, *J. Non-Cryst. Solids*, **134**, 183 (1991).
21. H. Kaji, K. Nakanishi, N. Soga, *J. Sol-Gel Sci. Technol.*, **1**, 35 (1993).
22. A.J. Vega, G.W. Scherer, *J. Non-Cryst. Solids*, **111**, 153 (1989).
23. P.J. Davis, C.J. Brinker, D.M. Smith, R.A. Assink. *J. Non-Cryst. Solids*, **142**, 197 (1992).
24. R.Y. Sheinfain, I.E. Neiwmark in *Absorption and Absorbents*, edited by D.N. Strazhesko, No. 1, p. 87–95. Wiley: N.Y. (1973).
25. B.S. Shukla, G.P. Johari, *J. Non-Cryst. Solids*, **101**, 263 (1988). D.M. Krol, J.G. van Lierop, *J. Non-Cryst. Solids*, **63**, 131 (1984).
26. J. Zarzycki, M. Prassas, J. Phallipou, *J. Mater. Sci.*, **17**, 3371 (1982).
27. G. Orcel, L. Hench, *J. Non-Cryst. Solids*, **79**, 177 (1986).
28. T. Adachi, S. Sakka, *J. Non Cryst. Solids*, **99**, 118 (1988). T. Adachi, S. Sakka, M. Okada, Y.K. Shi, *J. Non Cryst. Solids*, **95**, 970 (1987). T. Adachi, S. Sakka, *J. Non Cryst. Solids*, **100**, 250 (1988).
29. I. Artaki, T.W. Zerda, J. Jonas, *J. Non Cryst. Solids*, **81**, 381 (1986).
30. D.R. Ulrich *Ceramic Bull.*, **64**, 1444 (1985).
31. P.W. Atkins, *Physical Chemistry*, Oxford University Press (1978).
32. Japanese patent; *Chem. Abs.*, **111**, 27662 (1987).
33. K. Benner, P. Klüfers, J. Schumacher, *Angew. Chem. Int. Ed. Engl.*, **36**, 743 (1997).
34. J. Heilmann, W.F. Maier, *Angew. Chem. Int. Ed. Engl.*, **33**, 471 (1994).
35. C.T. Kresge, M.E. Leonowicz, W.J. Roth, J. Vartuli, J.S. Beck, *Nature*, **359**, 710 (1992).
36. A. Monnier, F. Schüth, Q. Ho, D. Kumar, D. Margolese, R.S. Maxwell, D.G. Stucky, M. Krishnamurty, P. Petroff, A. Fizouzi, M. Janicke, B.F. Chmelka, *Science*, **261**, 1299 (1993).
37. M. Ogawa, *J. Am. Chem. Soc.*, **116**, 7941 (1994).
38. S.J. Beck, J.C. Vartuli, W.J. Roth, M.E. Leonowicz, C.T. Kresge, K.D. Smitt, C. T.-W. Chu, D.H. Olson, E.W. Sheppard, S.B. McCullen, J.B. Higgins, J.L. Schlenker, *J. Am. Chem. Soc.*, **114**, 10834 (1992).
39. Q. Ho, D. Margolese, U. Ciesla, P. Feng, T.E. Gier, P. Sieger, R. Leon, P.M. Petroff, F. Schüth, D.G. Stucky, *Nature*, **368**, 317 (1994).
40. P.T. Tanev, T.J. Pinnavaia, *Science*, **267**, 865 (1995).
41. P.T. Tanev, T.J. Pinnavaia, *Science*, **271**, 1267 (1996). H. Yang, N. Coombs, G.A. Ozin, *Nature*, **386** (1997) 692.
42. P.T. Tanev, M. Chibwe, T.J. Pinnavaia, *Nature*, **368**, 321 (1994).
43. D.M. Antonelli, J.Y. Ying, Chem. Mater. **8**, 874 (1996).
44. Z.-R. Tian, W. Tong, J.-Y. Wang, N.-G. Duan, V.V. Krishan, S.L. Suib, *Science*, **276**, 926 (1997).
45. D.M. Antonelli, J.Y. Ying, *Angew. Chem. Int. Ed. Engl.*, **35**, 2014 (1996).
46. T. Abe, A. Taguchi, M. Iwamoto, *Chem. Mater.*, **7**, 1429 (1995).
47. C. Ottero et al. *Mater. Chem. Phys.*, **34**, 214 (1993).

48. M. Hartmann, S. Racouchot, C. Bischof, *Chem. Commun.*, 236 (1997).

49. D. Zhao, D. Goldfarb, *Chem. Mater.*, **8**, 2571 (1996).

50. S.A. Davis, S.L. Burkett, N.H. Mendelson, S. Mann, *Nature*, **385**, 420 (1997).

51. S. Mann, *J. Chem. Soc. Dalton Trans*, 3953 (1997).

52. H. Nakamura, Y. Matsui, *J. Am. Chem. Soc.*, **117**, 2651 (1995).

53. G.A. Ozin, *Acc. Chem. Res.*, **30**, 17 (1997).

54. R.C. Chambers, W.E. Jones, Jr., Y. Haruvy, S.E. Webber, M.A. Fox, *Chem. Mater.*, **5**, 1481 (1993).

55. M.J. van Bommel, T.N.M. Bernards, A.H. Boonstra, *J. Non-Cryst. Solids.*, **128**, 231 (1991).

56. Z. Zhang, Y. Tanigami, R. Terai, H. Wakabayashi, *J. Non-Cryst. Solids*, **189**, 212 (1995).

57. R.H. Baney, M. Itoh, A. Sakakibara, T. Suzuki, *Chem. Rev.*, 95, 1409 (1995).

58. D.A. Loy, K.J. Shea, *Chem. Rev.*, **95**, 1431 (1995).

59. J.L. Speier, *Adv. Organomet. Chem.*, **17**, 407 (1979).

60. A. Slama-Schwok, M. Ottolenghi, D. Avnir, *Nature*, **355**, 240 (1992).

61. A. Slama-Schwok, D. Avnir, M. Ottolenghi, *J. Am. Chem. Soc.*, **113**, 3984 (1991).

62. M. Prato, *J. Mater. Chem.*, 7 1097 (1997).

63. G. Carturan, R. Campostrini, S. Dire, V. Scardi, E. de Alteriis, *J. Mol. Cat.*, **57**, L13–16 (1989).

64. S. Braun, S. Rappoport, R. Zusman, D. Avnir, M. Ottolenghi, *Mater. Lett*, **10**, 1 (1990).

65. D. Avnir, S. Braun, M. Ottolenghi, *Supramolecular Architecture: Synthetic control in Thin Film and Surfaces*, edited by T. Bein, *Am. Chem. Soc.*, Vol. 499, p. 384. Washington D.C (1992).

66. L.M. Ellerby, C.R. Nishida, F. Nishida, S.A. Yamanaka, B. Dunn, J.S. Valentine, J.I. Zink, *Science*, **255**, 1113 (1992).

67. F. Akbarian, B.S. Dunn, J.I. Zink, *J. Phys. Chem.*, **99**, 3892 (1995).

68. B.M. Novak, *Adv. Mater*, **5**, 422 (1993).

69. G. Philipp, H. Schmidt, *J. Non-Cryst. Solids*, **63**, 283 (1984).

70. C.E. Fowler, S.L Burkett, S. Mann, *Chem. Commun*, 1796 (1997).

71. X. Feng, G.E. Fryxell, L-Q. Wang, A.Y. Kim, J. Lui, K.M. Kemmer, *Science*, **276**, 923 (1997).

72. P. van der Voort, E.F. Vansant, *J. Liq. Chrom. & Rel. Technol.*, **19**, 2723 (1996).

CHAPTER **4**

METAL OXIDE GELS

4.1 INTRODUCTION

Silica sol-gels have been chosen in the previous two chapters to illustrate the range of possible mechanistic and chemical studies in sol-gel science. However, the review of the historical development of the subject in Chapter 1 demonstrates that many of the earliest uses of the sol-gel process involved metal oxide gels rather than silica gels. This chapter will examine the chemistry of these materials, and their applications will be discussed in Chapter 7.

Metal oxides can be prepared by two classes of sol-gel methods depending on whether the precursor is a metal organic compound or an aqueous solution of an inorganic salt. In both cases the chemistry is dominated by the high electropositive character of metal atoms relative to that of silicon. For example, table 4.1 shows the estimated partial positive charge, $\delta(M)$, on the central atom in a series of metal ethoxides:[1]

TABLE 4.1

Ethoxide	$Zr(OEt)_4$	$Ti(OEt)_4$	$Nb(OEt)_5$	$Ta(OEt)_5$	$VO(OEt)_3$	$W(OEt)_6$	$Si(OEt)_4$
$\delta(M)$	+0.65	+0.63	+0.53	+0.49	+0.46	+0.43	+0.32

This means that the rate of nucleophilic attack on the central atom of metal alkoxides and related precursors is much faster than that on silicon alkoxide precursors (e.g. hydrolysis rate constants at pH 7 are 5×10^{-9} $M^{-1}s^{-1}$ for $Si(OEt)_4$ and 10^{-3} $M^{-1}s^{-1}$ for $Ti(OEt)_4$), and that hydrated metal ions have an increasing tendency toward acid dissociation as the electropositive character of the metal increases:

$$[M\cdots OH_2]^{z+} \rightleftharpoons [M-OH]^{(z-1)+} + H^+ \rightleftharpoons [M=O]^{(z-2)+} + 2H^+$$

Another important factor is the greater ability of metals to vary their coordination number and geometry depending on the size and charge of the ion, the number of d-electrons, the crystal field stabilisation energy and the nature of the surrounding ligands (as discussed in all basic inorganic chemistry texts). This provides great versatility of reaction mechanisms via a wide range of permissible transition state coordination geometries. In contrast the geometry of silicon is generally that of a tetrahedral covalent bonded species modified by the availability of vacant d orbitals and the reaction mechanisms are more limited, as already discussed in chapter 2.

These factors mean that great care must be taken when storing, handling and using metal organic precursors due to their high reactivity, particularly with moisture. They also lead to a very complex aqueous chemistry for many metal salts due to the series of possible equilibria involving acid dissociation and condensation reactions of the various hydrated species of various different oxidation states of the metal. The foundations for understanding the aqueous chemistry of metal ions were laid by Bjerrum,[2] Werner[3] and Pfeiffer[4] in 1907, which was before the development of the sol-gel method for production of, for example, new catalysts using these materials (1925, see chapter 1). However more detailed development of the ideas only occurred much later (Sillen,[5] 1959) and even quite recently in the partial charge model of Livage and co-workers.[1] Recent work increasingly illustrates the power and control of the sol-gel method applied to metal oxides in the light of this increased mechanistic understanding.

4.2 HYDROLYSIS AND CONDENSATION REACTIONS OF METAL SALT PRECURSORS

4.2.1 The Partial Charge Model

The key to understanding hydrolytic equilibria in aqueous solutions of metal salts lies in the calculation of the partial charges on the atoms. When two atoms combine, a partial electron transfer occurs until the electronegativity χ_i of each atom is equal to the mean electronegativity χ for the system. At this point the atoms have acquired appropriate partial charges δ_i. The electronegativity of an atom is related to its partial charge by the equation:

$$\chi_i = \chi_i^\circ + \eta_i \delta_i \tag{4.1}$$

where χ_i° is the electronegativity of the neutral atom and η_i is the "hardness", defined as;

$$\eta_i = k\sqrt{\chi_i^\circ}$$

where k is a constant (1.36 for electronegativities on the Pauling scale). $\qquad$ (4.2)

The mean electronegativity χ is given by

$$\chi = \frac{\sum_i p_i \sqrt{\chi_i^\circ} + kz}{\sum_i (p_i / \sqrt{\chi_i^\circ})}$$

(p_i is the stoichiometry of the ith atom in the compound and z is the net charge on the ionic species). $\qquad$ (4.3)

From these equations, the partial charge δ_i is given by

$$\delta_i = (\chi - \chi_i^\circ) / k\sqrt{\chi_i^\circ} \tag{4.4}$$

and can thus be calculated since the electronegativities of the neutral atoms, the stoichiometric composition and the ionic charge are all known. Although this procedure is tedious, by following it a great deal of quantitative insight into aqueous chemistry can be obtained, and this will now be illustrated with reference to results based on these equations. Although the reader may verify the numerical data by reference to the above equations if desired, the emphasis will be on the significance of the results rather than on the numerical procedures involved.

4.2.2 Equilibrium Species in Aqueous Solutions of Metal Salts

As mentioned earlier, the nature of the species present in aqueous solutions of metal salts is determined by the position of the equilibrium:

$$[M\cdots OH_2]^{z+} \rightleftharpoons [M\!-\!OH]^{(z-1)+} + H^+ \rightleftharpoons [M\!=\!O]^{(z-2)+} + 2H^+$$

The reaction $M^{\delta+}$-$O^{\delta-}$-$H^{\delta+}$ + $H_2O \rightleftharpoons M\text{-}O^- + H_3O^+$ will proceed to the right if the overall partial charge on OH is positive in the general species $[MO_NH_{2N-p}]^{(z-p)+}$, i.e. for all systems beyond the limiting case $\delta(OH) = 0$. Equating the overall ionic charge on the species to the sum of the partial atomic charges we may write:

$$\delta M + 2N\delta H + N\delta O - p\delta H = z - p \tag{4.5}$$

Hence,

$$p(1 - \delta H) = z - \delta M - 2N\delta H - N\delta O \tag{4.6}$$

But $\delta(OH) = 0$ for the limiting case, so this reduces to

$$p(1 - \delta H) = z - \delta M - N\delta H \tag{4.7}$$

$$\text{i.e. } p = (z - \delta M - N\delta H)/(1 - \delta H) \tag{4.8}$$

Then using the partial charge formula derived above (eqn. 4.4)

$$\delta_i = (\chi - \chi_i^\circ) / k\sqrt{\chi_i^\circ}$$

the value of p can be expressed in terms of the charge z, coordination number N and electronegativity χ_M° of the metal as follows (using the standard values of electronegativity of O and H):

$$p = 1.45z - 0.45N - 1.07(2.71 - \chi_M^\circ)/(\sqrt{\chi_M^\circ}) \tag{4.9}$$

This shows, as expected from chemical intuition, that proton removal is facilitated by high formal charge on the metal ion and, to a lesser extent, by low coordination number, with a relatively minor effect of electronegativity (increasing electronegativity from 1.3 to 1.8 only reduces p by 0.6). These effects can be seen in table 4.2 below, where predictions of this model $(2N - p$ values) are compared with the observed most acidic forms of the species. As can be seen, the predictions of the model agree with the observed species and equilibria. The lowest charged species (e.g. Ag^+, Mn^{2+}) show no acidic behaviour, and hydroxy species are only formed at high pH. The highest charged species (RuO_4) show no basic behaviour. Between these extremes species with varying numbers of O, OH and H_2O ligands are in equilibrium as shown in table 4.2.

TABLE 4.2

Metal	Formal charge	Coordination number N	Electro-negativity	$2N - p$	Observed Species
Ru	+8	4	1.78	−1.1	$[RuO_4]^0$
Mn	+7	4	1.63	0.5	$[MnO_4]^-/[MnO_3(OH)]^0$
Cr	+6	4	1.59	2.1	$[CrO_2(OH)_2]^0/[CrO(OH)_3]^+$
V	+5	6	1.56	8.4	$[VO_2(H_2O)_4]^+/[VO(OH)(H_2O)_4]^{2+}$
Ti	+4	6	1.32	10.2	$[Ti(O)(H_2O)_5]^{2+}/Ti(OH)(H_2O)_5]^{3+}$
Zr	+4	8	1.29	15.1	$Zr(OH)(H_2O)_7]^{3+}/[Zr(H_2O)_8]^{4+}$
Fe	+3	6	1.72	11.2	$[Fe(OH)(H_2O)_5]^{2+}/[Fe(H_2O)_6]^{3+}$
Mn	+2	6	1.63	12.7	$[Mn(H_2O)_6]^{2+}$
Ag	+1	2	1.68	4.3	$[Ag(H_2O)_2]^+$

A similar approach has been used to calculate the number of protons $(2N - q)$ that cannot be removed from the aqueous species even in the most basic conditions, with the results shown in table 4.3.

TABLE 4.3

Metal	z	N	$2N - q$	Observed species
Ru	+8	5	−0.5	$[RuO_5]^{2-}$
Mn	+7	4	−1.1	$[MnO_4]^-$
Cr	+6	4	0.2	$[CrO_4]^{2-}/[CrO_3(OH)]^-$
V	+5	4	1.4	$[VO_3(OH)]^{2-}/[VO_2(OH)_2]^-$
Ti, Zr	+4	5	5.0	$[MO(OH)_4]^{2-}/[M(OH)_5]^-$
Fe	+3	4	3.8	$[Fe(O(OH)_3]^{2-}/[Fe(OH)_4]^-$
Mn	+2	3	3.1	$[Mn(OH)_3]^-/[Mn(OH)_2(H_2O)]$
Ag	+1	2	2.3	$[Ag(OH)_2]^-/[Ag(OH)(H_2O)]$

4.2.3 Condensation and Polymerisation in Aqueous Solutions of Metal Salts

Although the above treatment only deals with the nature of the preferred monomeric species in solutions of metal salts, it leads directly to consideration of condensation processes and the formation of polymeric species, sols, gels and precipitates. In condensation, one of the ligands acts as an attacking group for linking with a second metal species and, depending on whether or not this species already has its preferred coordination number, one of the existing ligand groups may act as a leaving group. For pure aquo ions, water of hydration is a very poor nucleophilic attacking group and there is thus no tendency for condensation. For the pure oxo species, the highly negative partial charges on the oxygen atoms give very good nucleophilic attack properties, but the strong electrostatic attraction to the highly positive metal centre means that the oxygen atoms are very poor leaving groups. Condensation thus only occurs for oxo ions when the coordination shell can be expanded, permitting addition condensation.

4.2.3.1 Olation

The best species for promoting condensation reactions are the mixed hydroxo-oxo or hydroxo-aquo ions. In hydroxo-aquo species the hydroxo groups are effective for nucleophilic attack while water molecules are good leaving groups, and condensation occurs via "*olation*", with formation of hydroxy bridges, e.g.:

$$2[Cr(OH)(H_2O)_5]^{2+} \rightleftharpoons [(H_2O)_4Cr(OH)_2Cr(H_2O)_4]^{4+}$$

The rates of such processes depend on the size, charge and crystal field stabilisation energy of the metal ions, as discussed in all general inorganic chemistry texts; nevertheless, however rapid the reactions, they do not proceed indefinitely to form infinite polymers since the partial charge on the OH group(s) is changed in the bridging configuration. Thus in the above example the partial charges on OH in the monomer and dimer are -0.02 and $+0.01$ respectively, and the OH loses its nucleophilic property in the dimer. This does not always occur at the dimeric stage; for example in the following series of nickel complexes the partial charge on OH is -0.07 in the monomer, -0.03 in the dimer and $+0.06$ in the tetramer:

$$4[Ni(OH)(H_2O)_3]^+ \rightleftharpoons 2[(H_2O)_2Ni(OH)_2Ni(H_2O)_2]^{2+} + 2H_2O$$
$$\rightleftharpoons [Ni_4(OH)_4(H_2O)_4]^{4+} + 4H_2O$$

and the reaction therefore proceeds until the tetramer (figure 4.1) is formed.

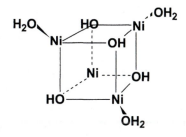

FIGURE 4.1 $[Ni_4(OH)_4(H_2O)_4]^{4+}$.

Similarly, for $[Zr(OH)_2(H_2O)_6]^{2+}$ with an initial OH partial charge of -0.07, a tetramer with Zr linked by double OH bridges (figure 4.2) is formed. In contrast, in some cases the partial charge on hydroxy groups is already very small even in the

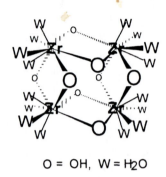

$$O = OH, \quad W = H_2O$$

FIGURE 4.2 $[Zr_4(OH)_8(H_2O)_{16}]^{8+}$.

monomer molecule and if there are two or more coordinated hydroxy groups in such systems, proton transfer is favoured rather than condensation, giving stable oxo cations. For example, in $[M(OH)_2(H_2O)_4]^{2+}$, with M = Ti or V, the OH partial charges are -0.01 and $+0.01$ respectively, and the stable equilibrium species are of the type $[MO(H_2O)_5]^{2+}$.

Since the hydroxy-aquo complexes discussed above form equilibrium mixtures with other species depending on pH, as shown in tables 4.2 and 4.3, the exact condensed species formed will also depend on pH, leading to a potentially very rich diversity of chemistry and a requirement for good control of pH if reproducible results are to be obtained.

4.2.3.2 Oxolation

Condensation may also occur by "*oxolation*", in which the metals are bridged by oxo groups rather than hydroxy groups. This may occur by direct addition reactions of coordinatively unsaturated oxo species (e.g. $4[MoO_3(OH)]^-$ react rapidly to form the cyclic tetramer $[Mo_4O_{12}(OH)_4]^{4-}$), or by nucleophilic addition followed by proton transfer and elimination of a leaving group:

$$M-\overset{\delta-}{OH} \; + \; \overset{\delta+}{M}-OH \xrightarrow[\text{catalyse}]{} M-O-M-\overset{\delta-}{OH} \xrightarrow[\text{catalyse}]{} M-O-M \; + \; H_2O$$

The initial nucleophilic attack is catalysed by bases, which remove a proton from the attacking OH group and increase its nucleophilic character. The second step is catalysed by acids, which protonate the leaving OH group. Hence the reaction occurs over a wide range of pH (unlike the olation reactions whose pH sensitivity is discussed above), and is slowest near the iso-electric point of the solution species. As with olation condensation, the extent of condensation is limited by the change in partial charge of the attacking OH groups as condensation proceeds. In the case of oxolation, however, the hydroxy groups in question are not the bridging groups – these are stable oxo bridges – but terminal groups. As their partial charge becomes positive, they become acidic and can undergo proton loss to form stable anionic species e.g. $[H_2Mo_{12}O_{40}]^{6-}$. Again, a very rich chemistry is opened up in this area, with a range of different polyanions for given metals, and the possibility of preparing mixed metal polyanions, polyanions with mixed-valence metals, and polyanions in which a central "hole" is filled with a hetero atom (e.g. phosphomolybdate $[PMo_{12}O_{40}]^{3-}$).

4.3   EFFECTS OF THE COUNTER-ION

The counter-ion has been ignored in the above discussions, but its effects are by no means always negligible. In chapter 1, the effects of counter-ions in altering the thickness of the double layer surrounding colloid particles and hence influencing coagulation was discussed. Small, highly-charged counter-ions promote coagulation and can thus exert a significant influence on the conversion of sol-gel materials into ceramic products.

Preferential adsorption of anions on different growing crystal faces may also modify crystal habits in the developing material, leading to a dependence of particle shape on the anion. This is well illustrated by the α-phase of the iron oxide Fe_2O_3, which forms small near-cubic prisms when the anion is Cl^- but elongated pointed needles when the anion is $H_2PO_4^-$.[6]

However, a further common effect when the counter-ion is negatively charged is the possibility of coordination to the metal centre and involvement in the solution chemistry, thus modifying the processes described earlier. Considering the case of monovalent anions interacting with a mixed hydroxy/aquo metal complex, two possible equilibria may exist. In the first, the anion displaces a water molecule from the coordination sphere of the metal, while in the second it displaces a hydroxy group:

$$[-M-X]^{(n-1)+} + H_2O \rightleftharpoons [-M-H_2O]^{n+} + X^-_{aq.}$$
$$[-M-X]^{n+} + H_2O \rightleftharpoons [-M-OH]^{n+} + HX_{aq.}$$

These reactions will be determined by the extent to which the anion (or the species HX) is a better nucleophilic electron donor than water to the positive metal centre, and by solvation energy and entropy terms. Anions which are good electron donors (i.e. which have low electronegativity) have a strong tendency toward protonation and thus undergo the second reaction unless the pH is very high. Anions which are poor electron donors (i.e. which are highly electronegative) remain as counter-ions, unless the pH is very low. Anions which lie between these extremes may become coordinated more easily over a wide pH range and then modify the hydrolysis and condensation reactions significantly. A way to predict the position of anions in this range is to compare their mean electronegativities with that of the water molecule (2.49). Thus, anions with low mean electronegativities (e.g. CH_3COO^-; 2.24) are easily protonated and do not coordinate easily. Ions with high mean electronegativities (e.g. ClO_4^-; 2.86) have a low tendency to coordinate and prefer to remain as the hydrated anion. Ions of intermediate mean electronegativity (e.g. HSO_4^-; 2.64) coordinate to the metal centre over a wide range of pH and are to be avoided unless their incorporation and the consequent modification of hydrolysis and condensation are deliberately used to control the characteristics of the end product.

4.4 REACTIONS OF METAL ALKOXIDE PRECURSORS

4.4.1 Hydrolysis and Condensation

When no acid or base catalyst is present, metal alkoxides react first by hydrolysis involving nucleophilic addition of a water molecule followed by proton transfer from water to the alkoxy group which then leaves as alcohol:

$$\begin{array}{c} H \\ \diagdown \\ O \\ \diagup \\ H \end{array} + M-OR \longrightarrow \begin{array}{c} H \\ \diagdown \\ O \rightarrow M-OR \\ \diagup \\ H \end{array} \longrightarrow HO-M \leftarrow O \begin{array}{c} R \\ \diagup \\ \diagdown \\ H \end{array} \longrightarrow MOH + ROH$$

This is followed by reaction of the resulting MOH species with a further alkoxide (alcoxolation):

$$\begin{array}{c} M \\ \diagdown \\ O \\ \diagup \\ H \end{array} + M-OR \longrightarrow \begin{array}{c} M \\ \diagdown \\ O \rightarrow M-OR \\ \diagup \\ H \end{array} \longrightarrow MO-M \leftarrow O \begin{array}{c} R \\ \diagup \\ \diagdown \\ H \end{array} \longrightarrow MOM + ROH$$

or another MOH species (oxolation):

$$\begin{array}{c} M \\ \diagdown \\ O \\ \diagup \\ H \end{array} + M-OH \longrightarrow \begin{array}{c} M \\ \diagdown \\ O \rightarrow M-OH \\ \diagup \\ H \end{array} \longrightarrow MO-M \leftarrow O \begin{array}{c} H \\ \diagup \\ \diagdown \\ H \end{array} \longrightarrow MOM + H_2O$$

or a solvated metal species (olation):

$$M-OH + M \leftarrow O \begin{array}{c} H \\ \diagup \\ \diagdown \\ R \end{array} \longrightarrow \begin{array}{c} M \\ \diagdown \\ O-H \\ \diagup \\ M \end{array} + ROH$$

$$M-OH + M \leftarrow O \begin{array}{c} H \\ \diagup \\ \diagdown \\ H \end{array} \longrightarrow \begin{array}{c} M \\ \diagdown \\ O-H \\ \diagup \\ M \end{array} + H_2O$$

The thermodynamics of these different processes are determined by the partial negative charge of the incoming nucleophile in hydrolysis, the partial positive charge of the electrophilic metal and the partial charge and stability of the leaving group (with more positively charged groups leaving most readily). Also, the initial attack of water on the alkoxide is easiest for alkoxides where the metal can easily expand its coordination sphere, and the proton transfer stage of hydrolysis oxolation and alcoxolation is clearly easier the more acidic the proton of the incoming coordinated species.

4.4.2 Acid and Base Catalysis

Similarly the effect of acid and base catalysts on the progress of condensation reactions can be predicted by calculating the charges on the species at various stages in the reaction. Thus for titanium alkoxide species the following charge distributions (δ) have been calculated:

TABLE 4.4

Species	$\delta(OR)$	$\delta(Ti)$
$Ti(OR)_3O-$	-0.08	$+0.68$
$Ti(OR)_2(OH)O-$	-0.01	$+0.70$
$Ti(OR)_2(O-)_2$	$+0.04$	$+0.71$
$Ti(OR)(O-)_3$	$+0.22$	$+0.76$

In acid catalysis, the least positively charged species will react fastest, i.e. the order of reactivity will be descending order in table 4.4, so chain end sites will be more reactive than chain centre sites, and long chains with little branching will be produced. Conversely for base catalysed reactions the most positively charged species will react fastest, i.e. in ascending order in table 4.4, so chain centre sites will be most reactive, leading to highly branched chains. This clearly provides control of the evolving oxide structure.

4.4.3 Steric Factors and Solvent Effects

Although these processes can be approached theoretically by calculations of charge distributions, such calculations do not necessarily provide valid predictions of reactivity, for three reasons:

(i) steric effects can also play a role;
(ii) because the oxidation state of the metal is often less than its normal coordination number, alkoxides are coordinatively unsaturated and oligomeric alkoxide species are often formed, whose steric and electronic properties are different from those of the parent monomers,
(iii) solvent interactions with leaving groups and with the initial alkoxide species can have substantial influences on the course of the reaction. For example, when $Zr(OPr^n)_4$ is dissolved in n-propanol, the propanol associates with the propoxide species and hydrolysis is rapid leading to a precipitate whereas, when the same precursor is dissolved in cyclohexane, alkoxy-bridged oligomeric metal alkoxide species are formed which hydrolyse more slowly leading to a gel rather than a precipitate.

Sometimes control over monomeric v. oligomeric precursors may also be achieved by using different alkoxy groups both in the alkoxide and the solvent alcohol. Thus tetra(ethoxy)titanium(IV) dissolves in ethanol to give an oligomeric species which hydrolyses in a controlled manner to produce monodisperse particles, whereas tetra(isopropoxy)titanium(IV) dissolves in isopropanol to give a monomeric species which hydrolyses rapidly giving a polydisperse precipitate. In some cases the formation of oligomers renders the alkoxides insoluble (e.g. alkoxides of Mn, Fe, Co, Ni, Cu). It is inappropriate in a text of this size to attempt a survey of the very wide range of metal alkoxides that are currently known, but a good introduction to the field is given in the book by Bradley et al.[7]

4.4.4 Control of Metal Alkoxide Reactions

The above examples illustrate the potential for controlling the reactions of metal alkoxides to produce metal oxide materials with desired characteristics. In addition to use of ligand and solvent to control formation of oligomeric precursor species, the alkoxide ligand itself can also be used to control reaction rates via both its electronic and steric influence. Thus, increasing the alkyl chain length increases the electron donating ability and reduces the charges on the metal and the transferred proton, as well as increasing the steric bulk of the alkoxy group. All of these factors slow both hydrolysis and condensation reactions, to the extent that for larger groups (e.g. tetra(n-butyloxy)titanium(IV) and tetra(t-amyloxy)titanium(IV)) stable sols are formed containing small species with limited extent of condensation. Phenoxy compounds are generally found to give slower hydrolysis than alkoxides, in marked contrast to the faster hydrolysis of silicon phenoxide than TEOS. In the silicon compound the negative inductive effect of the phenoxy group dominates, giving larger positive charge on Si and hence faster reaction. However for metals with d orbitals which can act as π-acceptors for the aromatic ring electrons of the phenoxy group, this effect dominates, reducing the positive charge on the metal and decreasing the reactivity. Finer control is achievable via the use of chelating ligands which form mixed complexes with alkoxide ligands, the chelating ligands being more inert towards displacement and thus providing control of the main condensation pathway.

A particularly elegant example is the use of acetylacetone to modify the reactivity of tetra(i-propoxy)titanium(IV). Addition of one mole of acetylacetone leads to a $Ti(OPr^i)_3acac$ complex whereas, with two moles of acetylacetone, $Ti(OPr^i)_2(acac)_2$ is formed. Using a combination of NMR, infra-red, XANES and EXAFS studies these complexes have been shown to have the following structures (figure 4.3):[8,9]

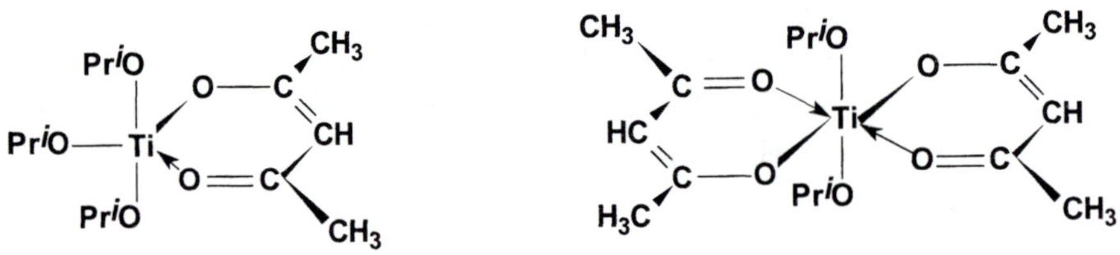

FIGURE 4.3 ACETYLACETONE COMPLEXES FROM TETRA(I-PROPOXY)TITANIUM(IV).

The alkoxy groups are hydrolysed more rapidly than the chelated acac groups, and on addition of a large excess of water a stable three-dimensional nanocrystalline material with crystal dimensions ~5nm is formed as a colloidal sol from the mono-acac complex[10] (c.f. 10–20nm for particles formed from hydrolysis of $Ti(OPr^i)_4$), whereas the bis-acac complex forms one-dimensional polymeric species which can be drawn into fibres.[11] Although care must be taken to ensure that all the chelating ligands are finally removed in applications where pure oxide material is required, this elegant approach clearly provides material with crystallite sizes and morphologies not easily obtainable by other means. Similar results have been obtained by treating metal alkoxides with acetic acid, which forms bidentate bridging carboxylate complexes.[9]

In addition to the above advantages of metal alkoxide precursors for sol-gel processing of metal oxides, it should also be noted that they can be obtained in very high purity by processes such as distillation and sublimation, and that alkoxides of different metals are often miscible, providing a route to controlled doping of metal oxides for electronic materials applications. However, they are often expensive and difficult to handle. It is therefore useful to note that nanocrystalline metal oxide materials may also be prepared from metal halide precursors either directly (e.g. from $SnCl_4.2H_2O$) or indirectly via reaction with alcohols to form chlorine-modified alkoxides which may then be hydrolysed. These routes are cheaper although probably do not afford the same purity and control as obtainable via the alkoxide precursors.

4.5   THE NON-HYDROLYTIC SOL-GEL METHOD FOR METAL OXIDES

The non-hydrolytic sol-gel process[12] discussed in chapter 2, section 2.4, is particularly useful for reactions which would otherwise involve precursors of very different reactivity. For example, silica-titania and silica-zirconia glasses,[13] where the much faster hydrolysis rates of the transition metal alkoxides leads to predominant formation of metal-oxygen-metal bonds and precipitation of the metal oxide. While this can be minimised if the metal alkoxide is stabilised by complexation as described elsewhere in this chapter, or if the silicon alkoxide is hydrolysed before the metal alkoxide is added, the non-hydrolytic method permits the formation of well condensed monolithic gels in one step without the use of any complexing agents or other additives. Furthermore, the absence of water means that no co-solvent is required.

This process is particularly useful for the preparation of mixed metal oxide systems normally requiring high-temperature processing to avoid decomposition with phase separation into the component oxides. For example, aluminium titanate Al_2TiO_5 is normally only stable above 1180°C, and conventional sol-gel processing leads to crystallisation of rutile TiO_2 between 710–770°C and α-Al_2O_3 between 910–1100°C. However, using mixtures of aluminium and titanium chlorides and alkoxides with Al:Ti ratios of 2 gave gels at 110°C which crystallised to pseudo-brookite (β-Al_2O_3) between 850–1000°C with no phase separation.[14] This is due to the very high homogeneity of the gels arising from the high rate of hetero-condensation in this non-hydrolytic method. Similar results have been obtained for other such systems e.g. $ZrTiO_4$.[15]

The mechanisms proposed for such reactions (figure 4.4) involve nucleophilic substitution of chloride on the carbon centre by S_N2 (figure 4.4a), concerted (figure 4.4b) or S_N1 (figure 4.4c) processes.

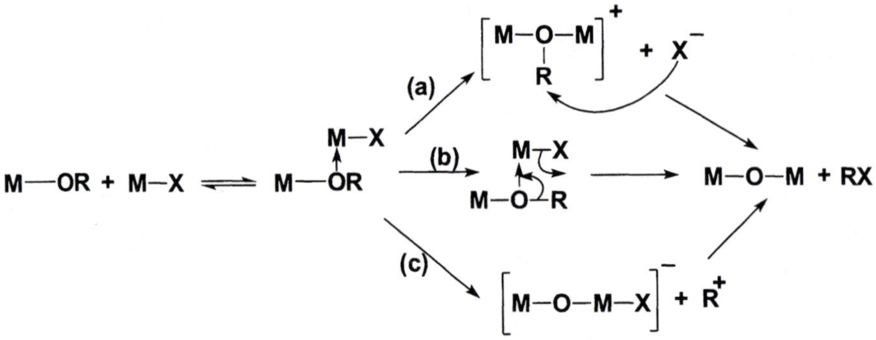

FIGURE 4.4 NON-HYDROLYTIC CONDENSATION MECHANISMS.

Product characterisation for reaction systems involving di-*n*-propyl ether as the oxygen donor show that the alkyl chloride is a 50:50 mixture of *iso*-propyl chloride and *n*-propyl chloride when the reaction proceeds in a sealed tube, but 20:80 when the alkyl chloride is removed by distillation as it is formed. Since the carbocation formed in the S_N1 mechanism (figure 4.4c) would rapidly rearrange to the more stable secondary cation yielding *iso*-propyl chloride, the observation that this only occurs as 20% of the distilled product argues against this mechanism playing a dominant role.

4.6 PARTICLE GROWTH AND AGGREGATION

The balance between van der Waals attractive and repulsive forces and the role of the thickness of the diffuse double layer in determining the stability of colloid particles against aggregation has been discussed briefly in chapter 1, and the description of particle growth in terms of the Smoluchowski equation has been outlined in chapter 2 (p.28). Although these two sets of ideas form the framework for understanding particle growth and aggregation, a closer look at this point is helpful in order to appreciate some of the approaches which have been adopted in order to prepare oxide and other materials with controlled and uniform particle sizes. This is a very important aspect in relation to the development of strong ceramics, since sintering of particle aggregates to form dense ceramic materials may lead to residual voids and defects if the initial pre-sinter particle packing is irregular. These voids may never close completely, and may form potential stress failure sites. Although uniform spherical particles may in principle pack perfectly and sinter with no voids, in practice some degree of "raft" formation is common, with spaces between the irregular perimeters of adjacent rafts of close-packed spherical particles. It has been argued that this problem is less serious if there is a slight dispersion in the particle sizes so that such voids may be filled more effectively. Clearly, control of particle size and uniformity is therefore very important for the development of improved ceramics.

The aggregation of colloidal particles produced in the stages described earlier in this chapter and in chapter 2 is determined by the detailed form of the attractive and repulsive potentials between neighbouring particles and by their modification by the effects of the diffuse double layer of ions around the particles. At the beginning of chapter 1 it was stated that the attractive van der Waals forces fall off as r^{-6} while electrostatic repulsions vary as r^{-1}. For real colloid particles this is only an approximation and, while it is inappropriate in a text of this level to venture into a discussion of more accurate potential functions, some conclusions based on improved functions are of interest. In general when conditions permit the diffuse double layer to be sufficiently thin the combination of attractive and repulsive curves leads to a secondary potential minimum, as shown in figure 4.5, with a potential barrier of height V_{max} barring closer approach. It has been shown that V_{max} is larger for larger

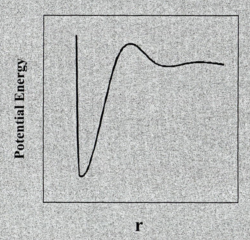

FIGURE 4.5 THIS CURVE WAS SIMULATED USING A WEIGHTED COMBINATION OF R^{-1} AND R^{-12} REPULSIVE POTENTIALS AND AN R^{-6} ATTRACTIVE POTENTIAL.

particles,[16] and that the effect of particle size on the potentials leads to possible aggregation of small particles with large particles even in conditions where small-small and large-large particle aggregation is not favoured.[17] For most aggregates of colloidal particles, it is this secondary potential minimum which determines the approach distance of the particles, and if the diffuse double layer can be re-established the precipitated colloid can be re-dispersed. This process is known as *peptization* and may be achieved by washing to remove the counter-ions which led to coagulation, or by adsorption of new ionic species to alter the nature of the charged surface and establish a new diffuse double layer. The Derjaguin-Landau-Verwey-Overbeek (DLVO) theory of electrostatic stabilisation of colloids[18-20] predicts that the critical ionic concentration to produce flocculation of a colloidal suspension should depend on

T^5/z^6, where T = absolute temperature and z = ionic charge. Thus, replacement of highly charged surface ions with ions of lower charge, and raising the temperature, helps peptization. This procedure has been shown to be useful in several cases where initial preparation of small metal oxide particles takes place in conditions where flocculation is favoured, and peptization followed by subsequent processing of the separated nanoparticles leads to more useful materials. In other cases, adsorption of polymeric or other bulky uncharged molecules may prevent the initial flocculation. (In the case of silica particles, the hydration sphere is in some cases sufficient to prevent coagulation even at the isoelectric point – the pH where there is no net effective surface charge.)

4.7 PREPARATION OF MONODISPERSE PARTICLES

Several general methods have been reported for the preparation of spherical particles of uniform size from solution:

i) Use of dilute (10^{-1}–10^{-3}M) solutions of metal salts in mild conditions with slow raising of pH (e.g. by thermal decomposition of formamide to produce ammonia).

ii) Thermal decomposition of complexes with chelating ligands in strongly basic solution. This leads to slow release of the metal ion as the chelating ligand dissociates, followed by immediate hydrolysis. It can be used for mixed oxide, for the production of particles of one material coated by another, and for production of small metal particles via incorporation of an in-situ reducing agent.

iii) Dispersion of an aqueous colloidal sol to a monodisperse emulsion in an immiscible organic solvent which can extract water from the sol (for example, 2-ethylhexanol). As water is lost, the emulsion droplets undergo gelation, forming uniform particles.

iv) Monodisperse silica particles (typically 95% within ±8% of the mean size) have been produced by mixing water (20-fold excess over TEOS), alcohol and ammonia and adding TEOS to form an opalescent suspension within 10 minutes.[21]

Originally it was thought that all of these methods depended on increasing the concentration until a critical nucleation concentration was reached, whereupon nucleation occurred reducing the concentration to below that required for further nucleation. Diffusive growth of the particles then occurred with no further nucleation until the solution reached its normal saturation concentration. However, experiments and associated calculations raised three major objections to this simple model:

a) Diffusion rates which fitted the observed growth kinetics were much slower than expected for such small particles.

b) The solution concentration was found to be high enough to allow further nucleation for a substantial part of the growth period.

c) Electron micrographs taken at various stages in the process showed small particles aggregating to form large uniform spherical agglomerates with rough surfaces.

It is therefore believed that aggregation of the small particles by a Smoluchowski mechanism (c.f. chapter 2) is the most likely underlying process for all these methods. The understanding of the ways in which inorganic materials from solution can assemble into an incredible variety of structures in mild conditions in biological organisms has greatly advanced in recent years via the study of *biomineralisation*. There is unfortunately insufficient space in such a text as this to embark on an outline of this area, but the interested reader is referred to a recent review and textbook.[22, 23]

REFERENCES

1. J. Livage, M. Henry and C. Sanchez, *Prog. Solid St. Chem.*, **18**, 259 (1988).
2. N. Bjerrum, *Z. Phys. Chem.*, **59**, 336 (1907).
3. A. Werner, *Ber.*, **40**, 272 (1907).
4. P. Pfeiffer, *Ber.*, **40**, 4036 (1907).
5. L.G. Sillen, *Quart. Rev.*, **13**, 146 (1959).
6. C. Sanchez, F. Babboneau, S. Doeuff and A. Leaustic, in *Ultrastructure Processing of Advanced Ceramics*, edited by J.D. Mackenzie and D.R. Ulrich. Wiley: New York, 1988.
7. D.C. Bradley, R.C. Mehrotra and D.P. Gaur, *Metal Alkoxides*. Academic Press: London, 1978.
8. J. Livage, C. Sanchez, M. Henry and S. Doeuff, *Solid State Ionics*, **32–33**, 633–638 (1989).
9. F. Babboneau, A. Leaustic and J. Livage, in *Better Ceramics Through Chemistry III*, Mater, Res. Soc. Symp. Proc. **121**, edited by C.J. Brinker, D.E. Clark and D.R. Ulrich, p. 317. Mater. Res. Soc.: Pittsburgh, Pa., 1988.
10. W.C. LaCourse and S. Kim, in *Science of Ceramic Chemical Processing*, edited by L.L. Hench and D.R. Ulrich, p. 310. Wiley: New York, 1986.
11. E. Matijevic, *Ann. Rev. Mater. Sci.*, **15**, 483 (1985).
12. R.J.P. Corriu, D. Leclercq, P. Lefèvre, P.H. Mutin and A. Vioux, *J. Mater. Chem.*, **2**, 673 (1992).
13. M. Andrianainarivelo, R. Corriu, D. Leclercq, P.H. Mutin and A. Vioux, *J. Mater. Chem.*, **6**, 1665 (1996).
14. M. Andrianainarivelo, R. Corriu, D. Leclercq, P.H. Mutin and A. Vioux, *Chem. Mater.*, **9**, 1098 (1997).
15. M. Andrianainarivelo, R. Corriu, D. Leclercq, P.H. Mutin and A. Vioux, *J. Mater. Chem.*, **7**, 279 (1997).
16. G.D. Parfitt, *Dispersion of Powders in Liquids*, p. 1–50. Applied Science: London, 1981.
17. E. Matijevic, in *Science of Ceramic Chemical Processing*, edited by L.L. Hench and D.R. Ulrich, p. 463–481. Wiley: New York, 1986.
18. B.V. Derjaguin and L. Landau, *Acta Physicochim.*, **14**, 633 (1941).
19. E.J.W. Verwey and J.T.G. Overbeek, *Theory of the Stability of Lyophilic Colloids*. Elsevier: Amsterdam, 1948.
20. The DLVO theory was developed independently by the two groups in references 18 and 19, working in Russia and the Netherlands, respectively. Due to the occupation of the Netherlands during World War II, contact between the two groups was restricted, leading to independent publication and incomplete cross-referencing, as recounted by the two groups at a Faraday Discussion in 1955 (*Disc. Faraday Soc.*, **18**, 180–181 (1955)).
21. W. Stöber, A. Fink and E. Bohn, *J. Colloid Interface Sci.*, **21**, 62–69 (1968).
22. S. Mann, *J. Mater. Chem.*, **5**, 935–946 (1995).
23. *Biomimetic Materials Chemistry*, edited by S. Mann, VCH: Weinheim, 1995.

CHAPTER 5

THE CHARACTERISATION OF SOL-GEL MATERIALS

5.1 INTRODUCTION

The analysis of sol-gel derived materials is a far from trivial task for two main reasons: the starting materials are very different from the final products (both chemically and physically), and secondly, due to the enormous versatility of the sol-gel method, the resulting materials (and their precursors) can cover a vast range of both physical and chemical properties. For example, the precursors are monomers and their reaction products polymers, the former generally are soluble, the latter insoluble. Furthermore, they may be hydrophobic vs. hydrophilic, liquid vs. solid, and during the process many intermediate stages are generated. The final products, however, may also be very different: they can be hydrophilic or hydrophobic, gels or glasses, films or bulk materials, inorganic or hybrid materials, dense or porous, tough or brittle. They can have various degrees of cross-linking and may be based on different metals such as silicon, titanium, vanadium and niobium, or mixtures of them. For these reasons a variety of chemical and physical characterisation techniques have been used to analyse the different aspects of these materials. The most important of these will be treated in this chapter. The chapter aims to convey the key features of the techniques with some representative examples, rather than attempting to give a comprehensive review of the contemporary literature in this vast area.

5.2 CHEMICAL CHARACTERISATION

5.2.1 Nuclear Magnetic Resonance

Using ^{29}Si NMR the chemical environment of silicon atoms in silicates can be analysed. For this both liquid and solid state NMR techniques are used, to probe the local chemistry of silicon before and after the sol-gel transition, respectively.

In silicates the Si-atoms are bound to four oxygen atoms and can be represented by a tetrahedron of which the corners link to other tetrahedra. In order to describe the substitution pattern around a specific silicon atom the Q^n notation is used in which Q represents a silicon atom surrounded by four oxygen atoms and n indicates the connectivity, i.e. the number of silicon atoms the Q unit is linked to.[1] Hence, Q^0 denotes monomeric units, Q^1 end groups, Q^2 middle groups in chains (or rings), Q^3 branching points and Q^4 fully interconnected silicate groups.

The peaks in NMR spectra are characterised by three main parameters: chemical shift, peak intensity, and line width. The chemical shifts found in the ^{29}Si NMR spectra of silicate range from −60 to −120 ppm. The monomeric Q^0 units give rise to a signal at the low field side of the spectrum. This signal shifts up-field by approximately 10 ppm for every Si-O-Si connection made to the central Si atom. The chemical shifts observed in the solid state correlate well with the chemical shifts observed in solution spectra for the same silicate structures.

In solid state NMR line broadening tends to increase going from Q^0 to Q^4, reflecting the diversity in the possible chemical environments of the different silicon atoms with equal n number.[2] Solid state NMR generally is not able to discriminate between species with different OR-groups (R = alkyl or hydrogen) on silicon. These differences can often be resolved using solution NMR, also due to the generally lower molecular weight of the material present before gelation.

Since the intensities of the signals generally are proportional to the number of corresponding atoms integration of the relative peak intensities give the distribution of the different structural sites. However, care has to be taken when cross polarisation (CP) is used.[3] The CP technique is used in solid state NMR in order to enhance signal intensity of nuclei with low gyromagnetic ratio γ (such as Si). The polarisation of low-γ nuclei is enhanced via polarisation transfer by dipolar coupling with high-γ nuclei (e.g. protons). Since the relaxation delay is no longer governed by the slowly relaxing Si nuclei, but rather by the much faster relaxation of protons, the delay between subsequent pulses can be decreased. Unfortunately, CP tends to overestimate the fraction of silanol-containing species, thereby underestimating the actual functionality of the materials. Furthermore, due to the fact that many Q^4 sites do not have protons in their proximity (distances <10Å) CP experiments only show a part of these sites and therefore cannot be used to obtain quantitative results.

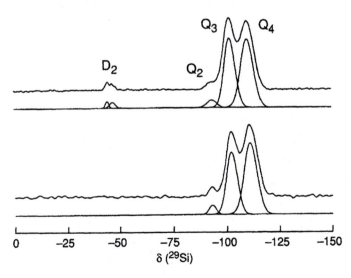

FIGURE 5.1 EXPERIMENTAL AND DECONVOLUTED ^{29}Si SOLID STATE NMR SPECTRA OF SOL-GEL GLASSES PREPARED FROM 100% TMOS (LOWER TRACE) AND FROM TMOS CONTAINING 4% OF OCTAPHENYLCYCLOTETRASILOXANE (UPPER TRACE).[4]

^{29}Si NMR has also been used to analyse ormosils generated by cocondensation of alkyltrialkoxysilanes (T^n) and/or dialkyldialkoxysilanes (D^n).[5] The signals corresponding to the respective T^n and D^n sites shift approximately 40 ppm down field for every alkoxy substituent that is replaced by an alkyl group (Figure 5.1). Using two dimensional ^{13}C/^{29}Si correlation spectroscopy the degree of mixing of Q^n and T^n sites can be determined.[6] In this way it is possible to discriminate between a cocondensate of MTES and TEOS, silica surface-modified with MTES, and a mixture of silica and polymethyl siloxane. Another way to determine the degree of mixing between different precursors is the use of ^{17}O-NMR. ^{17}O enriched sol-gel materials are easily prepared using enriched water for the hydrolysis.[7] In this way one can differentiate between domain formation, i.e. the formation of regions containing only Q^n, T^n or D^n sites and co-condensation with the formation of Si_Q-O-Si_T, Si_Q-O-Si_D or Si_T-O-Si_D linkages.[8] ^{17}O NMR can also be used to accurately determine the hydroxyl content of silicates.

5.2.2 Vibrational Spectroscopy

For the study of both the kinetics and the products of the sol-gel process vibrational (infrared and Raman) spectroscopy is very useful. Using infrared spectroscopy the hydrolysis and condensation of silicates can be conveniently monitored. Wood and Rabinovich showed that particularly the 700–1800 cm^{-1} range is useful to study the hydrolysis of TEOS.[9] Acid catalysis of the reaction leads to an instantaneous disappearance of the specific bands assigned to the backbone vibration of the precursor molecules (966 and 792 cm^{-1}) as well as the 1650 cm^{-1} vibration of water and the simultaneous appearance of the molecular backbone vibration of ethanol (880 cm^{-1}). At the same time Si-O-Si vibrations appear at 1160 and 1200 cm^{-1}, indicating the early stages of formation of a gel network. The development of the network can also be monitored by following the ν (Si-O-H) vibration (950–980 cm^{-1}) which decreases in intensity but also shifts to higher wavenumbers as the polycondensation proceeds.[10]

The wavelength at which the ν (Si-O-Si) skeletal vibration band appears in the FTIR spectrum depends on the degree of cross-linking of the silica network.[11] For highly cross-linked, base catalysed gels Si-O-Si vibrations have been observed at high frequency (1100 cm^{-1}) whereas for poorly cross-linked acid catalysed gels these vibrations were found at 1030 cm^{-1}.

The vibration at about 1200 cm^{-1} reflects a longitudinal mode and indicates the amount of long range order. The precise wavelength at which this vibration occurs depends on the magnitude of the depolarisation factor g_1. Since g_1 describes the shape of the silicate particles (0 for cylinders, 0.33 for spheres and 1.0 for plate-like structures) this vibration can be used to assess the microstructure of the material.[12]

Raman spectroscopy is an even more powerful method for the monitoring of the formation of intermediates during the sol-gel process. It can very accurately distinguish between the different substitution patterns of silicon as is demonstrated in Table 5.1. An illustrative example is the study of the post-gelation thermal treatment of sol-gel composites obtained from a mixture of TEOS and phenyltriethoxysilane.[13]

TABLE 5.1   ASSIGNMENT OF RAMAN SHIFTS FOR DIFFERENT INTERMEDIATES IN THE SOL-GEL PROCESS.

Intermediate	Raman shift (cm^{-1})	Assignment
$Si(OCH_3)_4$	644	Si-O-C
$Si(OCH_3)_3(OH)$	673	Si-O-C
$Si(OCH_3)_2(OH)_2$	696	Si-O-C
$Si(OCH_3)(OH)_3$	726	Si-O-C
$Si(OSi)(OR)_3$	608	
	586 (shoulder)	
$Si(OSi)_2(OR)_2$	525	
$Si(OSi)_3(OR)$	484	
$Si(OSi)_4$	432	Si-O-Si
dimers	795	Si-O-Si
silica network	830	Si-O-Si

5.3   PHYSICAL CHARACTERISATION

5.3.1   Nitrogen Adsorption Porosimetry

Langmuir described a method to determine the surface area of solids using the physical adsorption of inert gas molecules onto their surface.

The physical adsorption of gas molecules is driven by van der Waals-London forces. These forces are described by the term a/V^2 in the van der Waals equation:

$$(p + a/V^2)\,(V - b) = RT \qquad (5.1)$$

in which b is the volume occupied by the gas molecules and a/V^2 is the correction for the attraction forces between the gas molecules in which a is a constant characteristic for each gas.

If the adsorption of a gas is measured at a temperature well above the condensation temperature of that gas only a monolayer is formed, i.e. a second layer will not build onto the first one. By measuring the maximum amount of gas adsorbed the molecular area follows from

total surface area = (number of adsorbed gas molecules) × (area per molecule)

In the derivation of the isotherm describing monolayer adsorption, Langmuir used the following assumptions:

- the surface of the adsorbent is flat
- all adsorption sites are energetically equivalent
- the adsorbed gas molecules do not mutually interact
- the adsorbed molecules have a fixed position on the surface

In practice, however, these assumptions are seldom met: surfaces are never flat and adsorption sites are *not* energetically equivalent. Furthermore, adsorbed gas molecules *do* have mutual interactions, especially at higher surface coverages, and are highly mobile. Not taking into account surface heterogeneity causes deviations at relative pressures below $p/p_0 < 0.05$, whereas neglecting lateral interactions between the adsorbed gas molecules causes more serious mismatches at higher pressures ($p/p_0 < 0.35$).

Brunauer, Emmett and Teller developed a more practical model which still used the above-mentioned suppositions, but allowed for the adsorption of monolayers. In this model the assumption is made that adsorption forces are short range forces, i.e. that the heat of adsorption of the first layer is higher than that of following layers. In these other layers the heat of adsorption is assumed to be equal to the latent heat of condensation of the adsorbed gas.

The BET equation reads

$$\frac{p}{v(p^o - p)} = \frac{1}{v_m c} + \frac{(c-1)(p/p^o)}{v_m c} \tag{5.2}$$

where v is the volume of the gas adsorbed (at standard temperature and pressure (STP)) and v_m is the volume of gas (STP) adsorbed in the monolayer. c is equivalent to $\exp(Q-L)/R$, in which Q is the heat of adsorption of the first layer and L is the latent heat of condensation of the gas. p/p^o is the relative pressure of the gas.

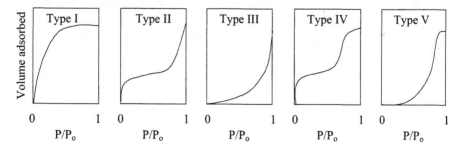

FIGURE 5.2 CHARACTERISTIC SHAPES OF THE FIVE CLASSES OF BET ADSORBTION ISOTHERMS.

Adsorption isotherms are plots of the amounts of gas adsorbed at equilibrium as a function of the partial pressure p/p^o at constant temperature, usually nitrogen at its boiling point (77.4 K). The obtained isotherms generally can be grouped in 5 classes, the characteristic features of which are depicted in figure 5.2. Isotherm type I is typical for adsorption in microporous materials (pore sizes < 2 nm) where the BET equation is not valid. Type II isotherms are characteristic for non-porous materials and types III and V are obtained for very weak adsorption interactions of which the fundamentals are not very well understood. For mesoporous materials isotherms of type IV are generally obtained. These isotherms show hysteresis in the adsorption-desorption cycles. This is explained by the Kelvin equation which says that the vapour pressure of curved surfaces differs from that of planar surfaces.[14] As a consequence the rate of desorption in pores can differ from the rate of adsorption, depending on the size and shape of the curve. This means that the shape of the isotherms can be related to the pore size distribution in the material. By applying the Kelvin equation in an appropriate form:

$$p/p_o = exp\ (-2\ \sigma V_L/(RTr)) \tag{5.3}$$

a graduated desorption experiment can be used to obtain a pore volume distribution in the form of a plot of $\Delta V^o/\Delta \bar{r}$ as a function of the average pore radius $\bar{r}$.

5.3.2 Mercury Porosimetry

Since nitrogen adsorption porosimetry cannot be used to analyse pore sizes larger than 20 nm,[15] it was suggested that the phenomenon of capillary depression could be used.[16] The minimum pressure p required to force a non-wetting liquid into a cylindrical pore of radius r is given by

$$p = -2\sigma\ cos\ \theta/r \tag{5.4}$$

where σ is the surface tension of the liquid and θ is the angle of contact between liquid and the pore surface. Since mercury has a very low wettability for most materials, pore size distributions in sol-gel materials can be determined by forcing this liquid into a porous specimen while measuring the amount of penetrating liquid as a function of the pressure p. Using the above-mentioned relation the pore radius r can be calculated and a pore volume distribution is obtained as a function of r.

This method in principle allows the determination of pore sizes ranging from 7.5–7500 nm.[17] However, using a model which assumes that the pores have a cylindrical shape may lead to the introduction of errors in the results. Pore sizes down to 2 nm have been measured using higher pressures,[18] although it should be noted that the validity of this method depends on the accuracy of the values that are used for the contact angle θ and the surface tension σ. These values may change when the liquid is confined in a small pore. Also the integrity of the pore shape is an issue of concern; under high pressures pore walls may flex or break leading to a change in porosity during the measurement. Nevertheless this method has successfully been compared to the nitrogen adsorption method for aluminosilicate samples (figure 5.3).

5.3.3 Thermoporosimetry

When water is confined in small pores a lowering of its melting point is observed, with the exception of a thin layer of non-freezable water which has very strong interactions to the pore wall. The melting point depression of the freezable part of the water can be related to the pore radius by means of the following equation:[19]

$$r = \alpha/\Delta T + \beta \qquad (5.5)$$

in which r is the pore radius, α is the proportionality coefficient that relates the freezing point depression of water to the pore size and β is the thickness of the layer of non-freezable pore water. For silica samples α and β were determined to be -64.67 nm K and 0.23 nm.[20]

Gibbs demonstrated that the melting transition temperature of pore material depends on the surface curvature. When the Gibbs equation on the thermodynamic stability of small crystals is combined with the Kelvin equation the Gibbs-Thompson equation is obtained which relates the decrease in melting point of a solid to its crystal dimensions in the following way: the temperature reduction ΔT_m of the melting point of a liquid confined in a pore in which a crystal is forming and of which the contact angles with the solid and pore wall are $180°$ is given by[21]

$$\Delta T_m = T_m - T_m(x) = 4sT_m/x\Delta H_f\rho \qquad (5.6)$$

where s is the surface energy at the liquid-solid interface, T_m the normal melting temperature, $T_m(x)$ the melting point of a linear dimension x, ΔH_f the bulk enthalpy of fusion and r the density of the solid. For a particular liquid this can be simplified to:

$$T_m - T_m(x) = k/x \qquad (5.7)$$

and since the pore volume $v(x)$ is a function of the diameter x, the pore size distribution follows from

$$dv/dx = kdv/(dT_m(x)\,x^2) \qquad (5.8)$$

Thermograms recorded in this way show differences between freezing and melting runs. These were initially attributed to the non-spherical character of the pores: the melting point depression depends on the curvature of the critical ice nucleus. In the case of a spherical pore the surface of the critical nucleus during solidification has the same curvature as during the melting. In the case of a cylindrical pore the freezing also starts from a spherical nucleus, however, the solid material adopts the shape of the pore and consequently the curvature of the solid-liquid interface will be different in the melting and freezing process. It was therefore proposed that this difference in the melting and freezing curves could be used to determine the shapes of pores.[22]

But nine years later it was later demonstrated by Ishikiriyama et al that the observed differences in freezing and melting curves were a consequence of the difference in temperature dependence of α upon freezing or melting and that the pore shape dependence was negligible.[23,24] They determined — for a pore shape in between spherical and cylindrical — that upon melting

$$\alpha(T) = 33.3 - 0.2\,\Delta T \qquad (5.9)$$

and upon freezing

$$\alpha(T) = 57.3 - 0.8 \, \Delta T. \tag{5.10}$$

On the other hand, β was found to vary from 1–4 monolayers depending on the pore shape.

5.3.4 NMR Spin-Spin Relaxation Measurements

The principle of melting point depression of confined liquids is also employed in the determination of pore size distributions using NMR.[25] The nuclear magnetic spin-spin relaxation time T_2 of a solid is typically short (typically 1 ms), whereas that of a liquid is longer (in the order of 1 s for cyclohexane). In a solid-liquid mixture one can differentiate between the solid and liquid component using a 90°-τ-180° pulse sequence with an interval of say $\tau = 20$ ms in order to make sure that only the long T_2 component contributes to the signal. Hence using this technique the melting of a fluid in a pore can be monitored by measuring the intensity of the proton signal which reflects the amount of liquid in the porous material. By measuring the signal strength as a function of the temperature the pore size distribution can be calculated using the Gibbs-Thompson equation in the same way as described for the thermoporosimetry measurements.

5.3.5 NMR Spin-Lattice Relaxation Measurements

The fluid in a porous material can be divided into a bulk fraction and a fraction that interacts with the pore wall. Using NMR techniques this difference can be used to determine pore size distributions in porous materials.[26] The presence of a pore wall increases the relaxation of a fluid inside these pores. Therefore, the spin lattice relaxation time T_1 of the confined fluid, typically water, will decrease with decreasing pore size.

In order to relate T_1 to the pore size, the "two-fraction, fast-exchange" model is used. This model assumes the exchange between the water bound to the pore wall and the bulk water to be much faster than the relaxation time. If this is the case than

$$1/T_1 = (f_{bulk}/T_{1bulk}) + (f_{surface}/T_{1surface}) \tag{5.11}$$

where f_{bulk} and $f_{surface}$ are the fractions of bulk and surface water. T_{1bulk} and $T_{1surface}$ are the relaxation rate constants for the bulk and surface phases and T_1 is the experimentally determined decay constant. Since the thickness of the surface layer is in the order of one monolayer $f_{bulk} \gg f_{surface}$ and the equation can be rewritten as

$$1/T_1 = \alpha + \beta/r_p \tag{5.12}$$

in which α is equal to $1/T_{1bulk}$, β equals $2/T_{1surface}$ and r_p is the pore radius. The method does not allow analysis of pore sizes smaller than approximately 4 nm due to the fact that in this situation $f_{bulk} \gg f_{surface}$ no longer applies.

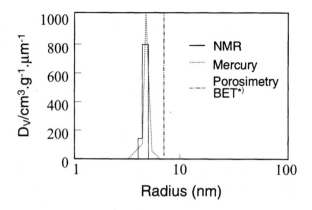

FIGURE 5.3 COMPARISON OF PORE SIZE ANALYSES USING SPIN-LATTICE RELAXATION NMR MEASUREMENTS, MERCURY POROSIMETRY AND BET-BASED MEAN PORE SIZE DETERMINATION OF A POROUS GLASS WITH A NARROW PORE SIZE DISTRIBUTION AROUND 7.4 NM.
) THE DEVIATION OF 2(PORE VOLUME/SURFACE AREA) FROM THE AVERAGE PORE SIZE IS PROBABLY DUE TO CONTACT ANGLE ERROR, SURFACE ROUGHNESS OR THE CYLINDRICAL PORE ASSUMPTION.[25]

After determination of α and β by calibration with standard materials T_1 measurements can be related to pore size. These T_1 distributions can be obtained from NMR measurements employing a specific pulse sequence (a $180° - \pi - 90°$ pulse sequence) and subsequent data analysis using the so-called non-negative least squares routine.[27]

The obtained pore size distributions correlate well with those from BET and Forced Mercury Porosimetry analyses (see figure 5.3). The main advantages of this technique are the fact that no pore shape assumptions are required but even more importantly that wet porous solids can be analysed directly without drying.

5.3.6 Small-Angle Scattering

The scattering of x-rays, neutrons and visible light by sol-gel samples provides a wealth of structural information. The nature of this information depends on the value of the scattering vector (variously given the symbols s, q or K) $(4\pi/\lambda)\sin(\theta/2)$, where θ is the scattering angle. (Note also that some authors replace $\theta/2$ by θ_{Bragg}, preferring to define the scattering angle as $2\theta_{Bragg}$.)

For large values of the scattering vector (to be termed q here), the distance d $(= 2\pi/q = \lambda/(2\sin\theta_{Bragg})$, as in Bragg's law) which is probed is very small and comparable with interatomic spacings. Since sol-gel materials are generally amorphous, no conventional sharp diffraction peaks are observed in this region, but diffuse scattering occurs from which radial distribution curves showing the distribution of near neighbour atoms may be derived.

At very small angles, Guinier[28] showed that the variation of scattered intensity ($I(q)$) with scattering vector followed the relation

$$I(q) \propto \exp(-R_g^2 q^2/3) \qquad\qquad (5.13)$$

where R_g is the radius of gyration. Hence for monodisperse systems a plot of $\ln I(q)$ v. q^2 will be linear with a slope $-R_g^2/3$, while for polydisperse systems such a plot will be curved. Since the radius of gyration is defined as $(I/m)^{1/2}$ where I = moment of inertia and m = particle mass, this can provide information on the distribution of mass within the particles in a sol-gel system. For example, the difference between the structures of acid and base catalysed gels (figure 2.6) appears as a complete independence of radius of gyration on dilution for base catalysed gels, in marked contrast to an apparent increase in radius of gyration on dilution for acid-catalysed gels. This has been explained in terms of entanglement of the more open polymer chains in more concentrated gels for the acid catalysed system. Upon dilution the chains disentangle, leading to a more extended conformation with larger radius of gyration, whereas the dense highly cross-linked base-catalysed gel particles do not entangle and hence show no dependence of radius of gyration on concentration.[29]

Between these two extremes the scattering intensity decays with increasing scattering vector following a power law, known as the Porod region:

$$I(q) \propto q^P$$

where P is a negative number known as the Porod slope (Figure 5.4). This region provides information on the nature of the internal structure of the sol-gel particles. Multi-functional centres such as silicon atoms with up to 4 polymer-bonding linkages form branching fractal structures.[30] Conventional crystalline solids with long-range order increase in mass directly in proportion to the volume of the sample, so for a spherical object of radius r the mass increases as r^3. However, random irreversible covalent branching, as in sol-gel silicates, leads to structures which become less dense as they grow (because of the greatly increased branching possibilities and associated steric constraints as size increases). In such a system (mass fractal), the mass increases as a smaller power of radius than 3, typically between 1 and 3, and so the Porod slope is -1 to -3. However, another possibility is that the branching occurs in such a way as to produce a structure composed of a very crumpled surface (like a sheet of paper pressed loosely into a ball). In this case the total surface area increases faster than that of a solid sphere (whose surface area would increase as r^2), in fact depending on r^3. In such a case the Porod slope lies between -3 and -4, and the corresponding structure is known as a surface fractal. The dimension of the surface fractal is given by $(6 - P)$, and P = 3 corresponds to a dense structure while P = 4 corresponds to smooth-surfaced fractals. Such studies on sol-gel derived silicas prepared from TEOS in a range of conditions (acid and base catalysed, with a range of r values) shows a wide range of behaviour, ranging from slopes of around -2 for mass fractal systems through quite dense mass fractal systems with slope close to −3 and on to surface fractal species with slopes ranging from just over −3 to −4 for smooth colloidal particles.

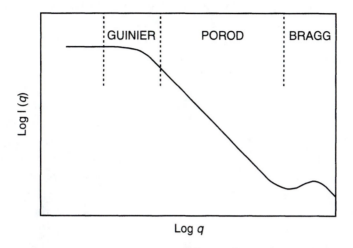

FIGURE 5.4   THE OVERALL SHAPE OF THE SCATTERING CURVE FOR A SOL-GEL MATERIAL.

It is possible to scan over the whole range of scattering vectors q to probe dimensions ranging from 0.1nm to 1µm by using a combination of light and x-ray scattering. The increasing availability of synchrotron radiation sources spanning this entire range with high intensity and excellent collimation is a valuable contribution to the growing popularity of this characterisation method. The use of neutron sources can give special advantages because neutron scattering can vary markedly for different atom types, and can even change sign when different isotopes are used, as shown in table 5.2.

TABLE 5.2   NEUTRON SCATTERING LENGTHS OF DIFFERENT ELEMENTS.

Element	O	H	D	Si	Ti	Zr
Neutron scattering length/10^{-12}cm	+0.577	−0.378	+0.65	+0.42	−0.34	+0.62

Thus, by varying the $H_2O:D_2O$ ratio in a series of titania/zirconia sols, the scattering from the water (H_2O/D_2O mixture) could be adjusted to be the same as that from the zirconia particles, giving minimum contrast. Hence either the scattering from both components or just that from the titania species could be studied to provide an improved model for the system.[31] Similarly, small angle neutron scattering allowed independent study of a mixture of spherical titania and rod-like FeOOH particle.[32] Clearly small-angle scattering experiments have great potential to reveal details of the structure of complex sol-gel materials. A strong advantage is that scattering experiments are non-invasive and there is no risk of irreversibly changing the sample (in contrast with the pore wall damage which can result from cryoporometry or forced mercury porometry), and that by conducting the experiment over a wide range of scattering vectors information on a correspondingly wide length scale may be obtained. However, the possible complications of polydisperse samples and the difficulty of interpreting Porod slopes in terms of unique material structures represent severe challenges to the future development of this technique and it is generally used in combination with a range of other characterisation methods.

5.3.7 Other Structural Techniques

In addition to their application in small-angle scattering studies, X-rays can also be used for crystallite size determination from **diffraction line-broadening studies**, and in studies of the local environments of specific atoms in amorphous materials via the technique of **extended X-ray absorption fine structure (EXAFS)** studies. Line-broadening occurs in the diffraction patterns from very small crystallites (e.g. nanocrystals) because of the finite size of the crystal lattice. According to the Bragg equation, x-rays are only scattered exactly in-phase from adjacent crystal planes when the Bragg angle exactly satisfies the Bragg equation $n\lambda = 2d\sin\theta$. For a slightly different angle, scattering from successive planes will suffer a small phase shift and, provided the crystal is large enough for the number of successive planes of a given type to be regarded as effectively infinite, destructive interference will occur giving a sharp line at the Bragg angle. However, in very small crystals this condition is not satisfied and for small deviations from the Bragg angle the interference is not complete and the diffraction peak has a finite linewidth. The Scherrer equation[33] gives the average crystallite size as $k\lambda/(\beta\cos\theta)$, where k is the Scherrer constant (which can range from 0.89–1.39), λ is the X-ray wavelength and β is the halfwidth of the line. Although this equation is approximate, and the halfwidth values require correction for instrumental broadening, the method is particularly valuable for continuous monitoring of changes in crystallite size during processing stages since it is rapid and non-invasive. In many cases, crystallite sizes determined in this way have been independently verified using electron microscopy, but the latter technique is virtually impossible to apply in situations where continuous monitoring of size is required.

In sol-gel chemistry, particles may grow without becoming crystalline and, even in cases where they are crystalline, line-broadening and disordered regions near the surfaces of very small crystals may greatly complicate detailed structural characterisation by conventional X-ray crystallography. In such circumstances, EXAFS studies[34] can be very valuable for characterising the local environments of specific types of atom and any changes in these environments during processing. When a sample is illuminated by X-ray photons of increasing energy, absorption increases very sharply at the X-ray absorption edge of each type of atom in the sample. Above this energy, photoelectrons are emitted from the atoms as spherical waves centred on the atom. Neighbouring atoms back-scatter the spherical waves, and interference occurs between the outgoing and returning wavefronts, leading to a modulation in the X-ray absorption coefficient whose detailed structure depends on the number, type and distance of each neighbouring atom and on the thermal motions of these atoms. Hence, if the modulations of the absorption coefficient are measured and these parameters are adjusted to give a best fit with the observed data, information on the local chemical environment of the absorbing atom is obtained. Using synchrotron radiation with a motor-driven double-crystal monochromator, EXAFS scans can be made in a few minutes. Furthermore, the technique does not require crystalline samples since it measures only local environments rather than long-range order.

A particularly valuable technique, often used in conjunction with small-angle scattering, is the direct imaging of sol-gel materials using **electron microscopy and atomic force microscopy**. A combination of scanning electron microscopy (SEM) and transmission electron microscopy (TEM) can give structural information over a very wide range of magnifications, with information on the chemical composition of the sample available via techniques such as **energy dispersive X-ray analysis (EDAX)** or **electron energy loss spectroscopy (EELS)**. Although the images are strictly 2-dimensional, examination of fractured samples can provide images of changing structure on progressing from the surface into the bulk of a sample, and this has been extensively used in the study of films and coatings. Since the sample must be held in a vacuum for electron microscopy, this technique is not suitable for the study of wet gels. Recently atomic force microscopy (AFM)[35] has begun to be used extensively for studying sol-gel samples, as it provides true three-dimensional images without the need to have the sample in vacuum, and at a significantly lower cost than for electron microscopy of comparable magnification. In this technique a fine needle probe mounted on a spring cantilever is scanned across the sample surface using two orthogonal piezoelectric ceramic drivers, while the distance between the probe tip and the sample surface is maintained constant using a third perpendicular driver (for example, by maintaining a constant spring cantilever deflection). The coordinates of the probe tip are recorded as a function of time, from the positions of the three drivers, giving a three-dimensional map of the surface down to atomic dimensions in favourable cases. Special variants involving intermittent contact between the tip and the surface ("tapping mode" AFM) have also been devised to overcome the problem of surface damage caused by the tip for soft samples. Although no chemical information is available from this method, it is very much easier to use at high magnification than is the case for electron microscopy, and can provide valuable information on particle sizes and surface physical structure.

Where information on the spatial variations in chemical composition in a sol-gel material is required, a number of techniques are available in addition to those of EDAX and EELS mentioned above. **Scanning Auger microscopy** can provide maps of the distribution of different elements in a sample. Surface-specific analytical methods such as **X-ray photoelectron spectroscopy** can be used together with etching to provide depth profiles of chemical composition in a sample. In such cases, care must be taken to ensure that the etching is uniform and does not itself influence the depth profile (for example, by preferential thermal diffusion of certain types of atom). **Rutherford backscattering spectroscopy (RBS)** has also been used to determine composition profiles in thin samples.[36] In this method, ions (typically He⁺) are accelerated using a Van der Graaff accelerator and fired at the sample in defined directions. The number and energy of backscattered ions is determined by the number and type of atoms in the material, respectively. By varying the angle of incidence of the ions, different depths from the sample surface are probed, providing information on the depth profiles of atom concentrations. As with electron microscopy, all of these experiments must be conducted under vacuum conditions. Full details of SEM, TEM, AFM and the methods for compositional analysis will be found in standard texts on these techniques, and further detailed discussion is inappropriate in a sol-gel text.

5.4 INDIRECT CHARACTERISATION METHODS

The changes in the vibronic finger print of emission spectra of pyrene were used to study the polarity changes during the polymerisation of silica sol-gel precursor mixtures. Pyranine (8-hydroxy-1,3,6-pyrenesulphonic acid) dissociates in aqueous media, and its anion fluoresces at a wavelength different from that of the undissociated acid. This made it possible to use this compound as a probe for monitoring the uptake and release of water molecules during the sol-gel process.[37]

ReCl(CO)$_3$-2,2'-bipyridine shows a **luminescence** band that shifts to higher energy when the molecule is constrained by its environment. This phenomenon can be used to follow the changes in viscosity during the development of the sol-gel network. It was found that gelation of silicates did not lead to rigidity at a molecular level; a rigidochromic effect was only observed in later stages of the drying process. When this compound is used to probe aluminosilicates, however, rigidity is demonstrated in much earlier stages of the sol-gel process, indicating early cage formation in this case.[38]

Microscopic viscosity can also be probed using **fluorescence polarisation** measurements. If molecules in an isotropic solution are excited using polarised light, the molecules whose absorption transition dipoles are aligned parallel to the excitation are selectively excited, and the resulting fluorescence will also be polarised. The fluorescence anisotropy **r** is defined as

$$r = \frac{I_{//} - GI_{\perp}}{I_{//} + 2GI_{\perp}} \tag{5.14}$$

where $I_{//}$ and $I_{\perp}$ are the fluorescence intensities of the vertically and horizontally polarised emissions respectively when the sample is excited with vertically polarised light, $(I_{//} + 2I_{\perp})$ is the total fluorescence intensity and G is a factor correcting for the polarisation characteristics of the monochromator. If the molecules rotate during the lifetime of the excited state, the anisotropy is reduced to

$$r = \frac{r_o}{1 + (\tau / \phi)}$$

(5.15)

where r_o is the anisotropy in the absence of molecular rotation, τ is the excited state lifetime and ϕ is the correlation time for rotation (given by $\phi = \eta V/kT$, where η is the viscosity and V is the volume of the rotating molecule). Hence measurements of r provide data on the microscopic viscosity of the immediate surroundings of the fluorophore, and can be used for studies of gelation and of the effect of pore size on molecular rotation.[39]

The degree of isolation of pyrene molecules could be used to identify isostructural silica gels.[40] The excimer/monomer ratio of these molecules typically changes from ~ 0.4 to ~ zero during the process.

The cage polarity of sol-gel glasses can be determined from spectral shifts of rhodamine 6G by comparing them with those obtained in various solvents.[41] The microenvironment of glasses can be more qualitatively assessed by measuring the Stokes shift of an incorporated solvatochromic dye, a measure for the orientational polarizability, which depends on the polarity of the matrix.[42] PRODAN (6-propionyl-2-(dimethylamino)naphthalene) was used to probe the interior of silica, PMMA and a silica/PMMA hybrid material. The results revealed that in the composite materials the dye molecules experience a multidomain microenvironment, in general very similar to a PMMA matrix but with a distinct influence of the silica matrix and the consequent hydrogen-bonding interactions. Based on this a model was proposed in which the composite consists of a silica matrix containing narrow channels filled with PMMA in which the dye molecules were immobilised.

REFERENCES

1. E. Lippmaa, M. Mägi, A. Samoson, G. Engelhardt, A.-R. Grimmer, *J. Am. Chem. Soc.*, 102, 4889 (1980).
2. G. Engelhardt, D. Michel, *High Resolution Solid State NMR of Silicates and Zeolites*. Wiley, NY, 1987.
3. M.P.J. Peeters, W.J.J. Wakelkamp, A.P.M. Kentgens, *J. Non-Cryst. Solids*, 189, 77 (1995).
4. N.A.J.M. Sommerdijk, E.R.H. van Eck, J.D. Wright, *Chem. Commun*, 159 (1997).
5. R.H. Glaser, G. L. Wilkes, C.E. Bronnimann, *J. Non-Cryst. Solids*, 113, 73 (1989).
6. C.A. Fyfe, Y. Zhang, P. Aroca, *J. Am. Chem. Soc*, 114, 3252 (1992).
7. F. Babonneau, J. Maquet, J. Livage, *Chem. Mater.*, 7, 1050 (1995).
8. F. Babboneau, C. Toutou, S. Gaveriaux, *J. Sol-Gel Sci. Technol.*, 8, 553 (1997).
9. D.L. Wood, E.M. Rabinovich, *J. Non-Cryst. Solids*, 107, 199 (1989).
10. D. Niznanski, J.L. Rehspringer, *J. Non-Cryst. Solids*, 180, 191 (1995).
11. M. Schraml-Marth, K.L. Walther, A. Wokaun, B.E. Handy, A. Baiker, *J. Non-Cryst. Solids*, 143, 93 (1992).
12. M. Ocana, F. Fornes, C.J. Serna, *J. Non-Cryst. Solids*, 107, 187 (1989).
13. J.C. Panitz, A. Wokaun, *J. Sol-Gel Sci. Technol.*, 9, 251 (1997).
14. P.W. Atkins, *Physical Chemistry*. Oxford University Press: UK, 1972.
15. E.P. Barret, L.G. Joyner, P.P. Halena, *J. Am. Chem. Soc.*, 73, 373 (1951).
16. W. Washburn, *Phys. Rev.*, 17, 273 (1921). E.W. Washburn, E.W. Bunting, *J. Am. Ceram. Soc.*, 5, 48 (1922).
17. L.C. Drake, H.L. Ritter, *Ind. Eng. Chem. Anal. Edn.*, 17, 787 (1945).
18. L.C. Drake, *Ind. Eng. Chem. Anal. Edn.*, 20, 780 (1949).
19. M. Brun, A. Lallemand, J.F. Quinson, Ch. Euraud, *Thermochimica Acta*, 21, 59 (1977).
20. J.F. Quinson, M. Astier, M. Brun, *Applied Catal.*, 30, 123 (1987).
21. C.L. Jackson, G.B. McKenna, *J. Chem. Phys.*, 93, 9002 (1990).
22. J. Quinson, J. Dumas, J. Serughetti, *J. Non-Cryst. Solids*, 79, 397 (1986).
23. K. Ishikiriyama, M. Todoki, K. Motomura, *J. Colloid Interface Sci*, 171, 92 (1995).

24. K. Ishikiriyama, M. Todoki, *J. Colloid Interface Sci*, **171**, 103 (1995).

25. J.H. Strange, M. Rahman, *Phys. Rev. Lett.*, **71**, 3589 (1993).

26. D.P. Gallegos, K. Munn, D.M. Smith, D.L. Stermer, *J. Colloid Interface Sci.*, **119**, 127 (1987).

27. K. Munn, D.M. Smith, *J. Colloid Interface Sci.*, **119**, 117 (1987) .

28. A. Guinier and G. Fournet, *Small Angle X-ray Scattering*. Wiley: London, 1955.

29. D.W. Schaefer and K.D. Keefer in *Better Ceramics Through Chemistry*, edited by C.J. Brinker, D.E. Clark and D.R. Ulrich, p. 1–14. Elsevier North Holland: New York, 1984.

30. J.E. Martin and A.J. Hurd, *J. Appl. Cryst.*, **20**, 61 (1987).

31. J.R. Bartlett, D. Gazeau, Th. Zemb and J.L. Woolfrey, *J. Sol-Gel Sci. Technol.*, **13**, 113 (1998).

32. J.D.F. Ramsay, R.G. Avery and L. Benest, *Faraday Disc. Chem. Soc.*, **76**, 53 (1983).

33. H.P. Klug and L.E. Alexander, *"X-ray diffraction procedures for polycrystalline and amorphous materials"*. Wiley: New York, 1974.

34. B.K. Teo and D.C. Joy, *"EXAFS spectroscopy: Techniques and Applications"*. Plenum: New York, 1981.

35. R. Wiesendanger, *"Scanning probe microscopy and spectroscopy: Methods and Applications"*. CUP: Cambridge, 1994.

36. M.J. Martin, M.L. Calzada, J. Mendola, M.F. da Silva and J.C. Soares, *J. Sol-Gel Sci. Technol.*, **13**, 843 (1998).

37. V.R. Kaufmann, D. Avnir, D. Pines-Rojanski, D. Huppert, *J. Non-Cryst. Solids*, **99**, 378 (1989).

38. J. McKiernan, J.C. Pouxviel, B. Dunn, J.I. Zink, *J. Phys. Chem.*, **93**, 2129 (1989).

39. F.J. Perrin, *Phys. Radium*, **7**, 390 (1926).

40. V.R. Kaufmann, D. Avnir, *Langmuir*, **2**, 717 (1986).

41. D. Avnir, D. Levy, R. Reisfeld, *J. Phys. Chem.*, **88**, 5956 (1984).

42. R. Gvishi, U. Narang, F.V. Bright, P.N. Prasad, *Chem. Mater*, **7**, 1703 (1995).

CHAPTER 6

APPLICATIONS OF SOL-GEL SILICATES

6.1 INTRODUCTION

Sol-gel technology provides an alternative route to the production of ceramics and glasses. Compared to conventional techniques the sol-gel route offers a number of important advantages that make the method interesting for the production of materials tailored to specific applications. The main potential lies in the fact that non-metallic, inorganic solids can be produced and processed at temperatures which are considerably lower than those required in conventional methods. This allows the incorporation of organic molecules and polymers, leading to materials with added functionality which cannot be obtained otherwise. Another advantage is the fact that these materials are obtained from solution which allows the convenient production of films and bulk materials of any possible shape. The third major advantage of the sol-gel method is that it produces porous materials whose pore-size distributions can be controlled, both by the chemical composition of the starting material, as well as by the processing conditions. The added value of the sol-gel approach for a number of applications will be illustrated using some selected examples highlighting the specific characteristics of this technology.

6.2 OPTICAL MATERIALS

6.2.1 Non-Doped Glasses

Modern communication and information technologies rely more and more on optical signal transduction using laser technology, optical fibers, non-linear optics (NLO) and many other high-tech components. The sol-gel method has been extensively explored as a low temperature route to such optical components.[1,2] One of the important aspects of the sol-gel method is the fact that glass components can be moulded into a large variety of shapes, which is particularly useful for the production of lenses. However, in the production of bulk pieces of optical quality glass, several problems were encountered including cracking upon drying of the gel, bloating during the gel-glass transition, and the presence of micro-cracks in the consolidated gels. Toki and co-workers demonstrated that by mixing colloidal silica into a hydrolysing TEOS solution the porosity can be increased such that capillary stress is significantly decreased during the drying and sintering processes. Using this method they were able to produce large pieces of crack free glass.[3]

The formation of cracks in sol-gel materials during drying is an even bigger problem in the production of optical quality thin films in which the whole volume contraction of the gel (200–600%) is restricted to one dimension only. In very thin films (0.5 μm) the stress generated by one dimensional shrinkage is still of low enough magnitude not to cause cracking, i.e. the energy required to extend a crack is greater than the energy gained from relief of stress near the crack. However, as soon as the thickness increases beyond a critical thickness this balance changes and cracking is very difficult to avoid.[4] Since cracking occurs due to stress arising from the volume contraction during the drying process, this problem can be solved by the reduction of the volume of the reactants and by avoiding a high degree of cross-linking in the silica polymer.[5] As a first step Haruvy and Webber omitted the solvent and used the minimal amount of water required to prevent precipitation of the TEOS (5:1, v/v). Secondly, they changed TEOS for TMOS which leads to a lower molar volume (149 vs 223 cm^3) with the additional advantage that less water is required to prevent the precipitation of a TMOS derived system (1:1, v/v). These measures led to a 40% lower volume reduction upon drying. In order to allow better stress relaxation they replaced TMOS by methyltrimethoxy silane (MTMOS) in order to block one direction of cross linking. In addition, the hydrolysis and condensation reactions were performed at elevated temperature (60°C) which promotes the formation of linear polymeric strands instead of cross-linking. The higher temperature also promotes the evaporation of the methanol released during the process, which leads to a reduction of the reaction volume in the early stages of the process. This procedure is generally applicable when a quick low temperature route to high quality films is required.

Ultrasound can be used to mix silicon alkoxides and water without the presence of a solvent, yielding a clear solution.[6] In this procedure bubbles are generated in which the alkoxide and water react in the vapour phase. These solutions quickly gel to form so-called *sonogels* which are unique in the way that they possess a very fine and uniform porosity, and a high apparent density.[7] The sonication treatment is also very suitable for the molecular dispersion of dopants, and sonogels thus have been used for example in the fabrication of non linear optical (NLO) materials.[8] For this type of application optical quality surfaces are required in order to obtain reliable properties.[9] This could be achieved by sealing the pores through impregnating these materials with a freshly prepared *sonosol*. This reinforces the materials after which they can be polished to yield excellent optical properties.

In applications where delicate components are used, e.g. fibre optics, oxynitride glasses are preferred over the corresponding oxide glasses since they have better mechanical, chemical and thermal properties. These oxynitride glasses are difficult to prepare by conventional methods due to the high temperatures required. In the case of fibres the nitrogen content was limited to approximately 8% due to the difficulty in drawing of this material. However, fibres can be conveniently drawn from a sol-gel mixture when low amounts of water are used in an acid catalysed hydrolysis reaction.[10] Chemisorption of ammonia in these porous fibres and subsequent heat treatment (600–1400°C) yields silicon oxynitride fibres with nitrogen contents up to 32%.[11]

Waveguide gratings are used as planar waveguide couplers, narrow band filters, beam splitters, and focussing elements in integrated optics. These gratings are typically produced by electron beam writing or holographic methods. Using sol-gel techniques they can be embossed, i.e. imprinted by pressing a surface relief pattern in the gel, before cross-linking is complete (Figure 6.1). Since this technique offers a practical non-vacuum method for the large scale production of these gratings, much research has been focussed on its development.[12] Especially the shrinkage of the pattern during drying and the resulting grating depth and period have been studied. The incorporation of polyethylene glycol (PEG) into the sol-gel mixture can be used to control the viscosity of the sol and the hardness of the gel films.[13] Higher PEG concentrations delay the gelation and allow more time for embossing. In order to control the accuracy of the moulding stage the degree of cross linking at the imprinting stage can be fine-tuned via a pre-embossing heat treatment of the gel-film.[14] Furthermore, the characteristics of the waveguide strongly depend on the refractive index, and hence on the porosity of the final material. By changing the hydrolysis and condensation reactions as described in Chapter 3, sol-gel chemistry can be used to control the characteristics of the waveguides produced in this way.

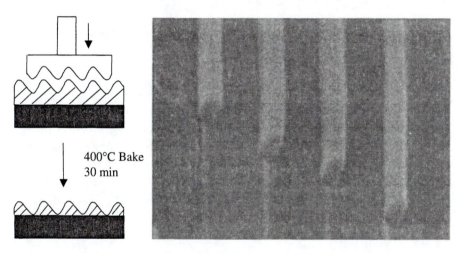

400°C Bake
30 min

FIGURE 6.1   THE PRODUCTION OF GRATINGS USING AN EMBOSSING PROCESS.[14]

6.2.2   Doped glasses

The most important advantage of the sol-gel technology in the field of optical materials is the fact that it permits the inclusion of organic compounds in the final glass. This allows for a great variety of functions including laser action,[15] photo-luminescence, photochemical hole burning,[16] photochromism,[17] NLO behaviour[18,19] and contrast enhancement. This section will illustrate the specific properties of the sol-gel matrix which enhance the functionality of the material.

Most organic molecules that are used for optical purposes are subject to photodegradation. Avnir et al. have demonstrated that encapsulation in a sol-gel matrix stabilises dye molecules,[20] most probably by preventing them from agglomeration. The fact that molecules can be isolated when entrapped in a silicate matrix sometimes also has advantages in terms of their functionality. For example the monocation of Oxazine-170 has excellent laser properties but readily forms a dimer which has only poor photoactivity. However, when trapped in a sol-gel matrix no dimerisation is observed. Moreover, the extinction coefficient of the dye in the glass can be optimised by changing the preparation procedure of the composite.[21]

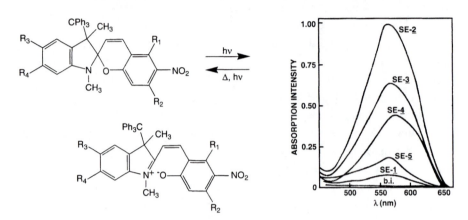

FIGURE 6.2 THE PHOTOCHEMICAL TRANSFORMATION OF SPIROPYRANES INTO THE OPEN ZWITTERIONIC COLOURED MEROCYANINES (LEFT) AND THE ABSORPTION SPECTRA OF SOME SILANE-ETHYL (SE) MODIFIED SPIROPYRANE-CONTAINING GLASSES AFTER IRRADIATION (RIGHT). THE COLOURLESS BASELINE OBTAINED BEFORE IRRADIATION (TYPICAL FOR ALL DYES) IS ALSO SHOWN.

In other cases the choice of the sol-gel precursor can be used to tune the function of the encapsulated compounds. For example spiropyranes encapsulated in a sol-gel matrix show photochromic behaviour (Figure 6.2); however, during the gel-xerogel transition the photochromism changed to reversed photochromism and eventually completely disappeared in the final xerogel stage. This phenomenon was attributed to the polarity of the environment and the decreased rotational freedom of the dye in the sol-gel cage. Ethyltrimethoxysilane-based gels which have limited cross-linking capacity showed normal photochromism indicating that the dye in this case indeed is entrapped in an apolar environment in which the matrix — due to its restricted cross-linking possibilities compared to TMOS — allows sufficient mobility for photochromic behaviour. Gels prepared from tetramethoxysilane and poly(dimethylsiloxane) (PDMS) gave rise to reverse photochromism. It was assumed that the incorporation of the linear PDMS strands prevented extensive cross-linking of the TMOS leading to a cage which was flexible enough to permit photochromism. However, in this case the dye is entrapped in a silica cage, *i.e.* a polar medium, causing reversed photochromic behaviour.

The ability to form hybrid interpenetrating organic-inorganic composites was utilised in the preparation of NLO materials based on conjugated polymers. Poly-*p*-phenylene vinylene (PPV) is a rigid rod-like polymer with a large third order susceptibility $\chi^{(3)}$ but is poorly soluble and its films have poor optical qualities. However, optical quality films could be prepared by dissolving the PPV precursor in a hydrolysing TMOS solution and spin coating this mixture. The PPV is then formed by thermal elimination reaction leading to an encapsulated polymer with an effective π-conjugation comparable with that of the neat polymer. The resulting composite had excellent non-linear properties and low optical loss.[18]

6.2.3 Contact Lenses

Materials for hard contact lenses must meet a variety of property requirements concerning flexibility, hardness, scratch resistance, refractive index, transparency, chemical stability, wettability and biological inertia. From silicone rubbers it was known that the dimethylsiloxane backbone provides high oxygen permeability. However, a more hydrophilic material was required to achieve wettability requirements. Si-OH groups could not be used since they were known to lead to protein deposition. Instead epoxy silicates were used since they hydrolyse to form a low density glycol-containing silica network with good wettability. Mechanical properties could be improved by the incorporation of $Ti(OR)_4$, thereby increasing the degree of cross-linking in the network ($Si(OR)_4$ could not be used due to its low reactivity). The resulting materials had excellent permeability, wettability and optical properties, but lacked sufficient flexibility due to the ionic character of the gel. This was solved by addition of methacryloxy silane to the sol-gel mixture which, after a pre-condensation step, yielded a viscous fluid. This was mixed with methyl methacrylic acid (MMA) and hydroxyethyl methacrylic acid (HEMA) and subsequently polymerised to give a material with good mechanical, biomedical and optical properties as indicated in Table 6.1.[22]

TABLE 6.1   MAIN PROPERTIES OF COMPOSITES[a] CONTAINING 20–30% MOL% METHACRYLATES[b].

Tensile strength[c]	4.85–5.15 MNm^{-2}
Modulus of elasticity	33–34 MNm^{-2}
Mohs' hardness	3
Refractive index[d] n_D^{20}	1.499–1.503
Contact angle with water[e]	25 $\pm$ 5°
O_2-permeability coefficient $P \times 10^{11}$	11.5–13.3 ml O_2 cm^2 ml^{-1}s^{-1} mm Hg

[a] Polymer composition: Epoxy silane, 1.5 mol% $Ti(OR)_4$ and 5 mol% methacryloxysilane.
[b] Methacrylates: MMA and HEMA. [c] Drafting speed 5 cm.min^{-1}. [d] in the dehydrated state.
[e] in the hydrated state.

6.3   CHEMICAL SENSORS

6.3.1   Optical Chemical Sensing

After the finding that organic molecules can be entrapped in a sol-gel matrix while still being accessible from solution it was soon realised that this concept offers possibilities for the construction of chemical sensors (figure 6.3). Zusman et al. demonstrated that entrapped analytical reagents still produce a response upon exposure of the composite to solutions of the analyte.[23] By using optical absorbance they detected cationic, anionic and organic species and showed that low detection limits (concentrations down to 10^{-7}M were detected) and fast response times (a few seconds) could be obtained. The main problem that was encountered was the diffusion of the reagents out of the silica matrix with time. Although leaching is not so much a problem for disposable sensors, for continuous monitoring the practicality of the method strongly depends on the efficiency of entrapment. In this case the problem of leaching can be overcome by covalently attaching the reagents to the silica matrix by equipping them with trialkoxysilane substituents.[24] The sol-gel method has been demonstrated to be an excellent method for the immobilisation of complexation reagents on optical fibres, thereby significantly extending the applicability of this technology.[25]

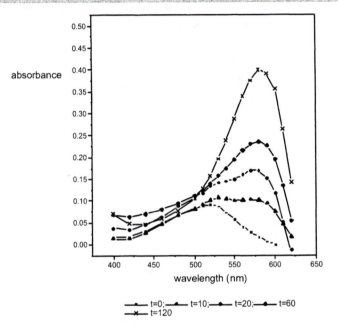

FIGURE 6.3 TIME RESOLVED RESPONSE IN MINUTES OF AN OPTICAL CHEMICAL SENSOR FOR CU(II) IONS BASED ON A SOL-GEL COMPOSITE CONTAINING ERIOCHROME CYANINE R.[26]

6.3.2 Biosensors

The highly selective and extremely efficient reactions of enzymes make them desirable reagents for sensing purposes but their performance is very sensitive to changes in their environment such as changes in pH, the presence of organic solvents, aggregation and immobilisation. However, researchers have been able to develop methods of immobilising enzymes in sol-gel materials without the loss of their function (see also chapter 3). By providing an enclosed compatible environment the sol-gel matrix protects the enzyme from aggregation, microbial attack, and denaturation. Another advantage is that enzymes, due to their size, do not leach from the sol-gel composite. Encapsulated metallo-enzymes have been used for the direct optical detection of O_2, CO and NO.[27] For the detection of glucose and oxalate, glucose oxidase (GOx) and oxalate oxidase were used, respectively. The oxidation of these analytes leads to the production of hydrogen peroxide but does not give rise to a colour change. However, the co-encapsulation of a mediator, e.g. horseradish peroxidase, could be used to generate an indirect colour production upon oxidation of the analyte.[28] Due to their low electronic conductivity, sol-gel silicates are not suitable for the transduction of electrochemical signals produced by redox enzymes. Sanchez and co-workers have demonstrated the generation of catalytic currents in the presence of glucose by using immobilised GOx with hydroxymethyl ferrocene as the mediator.[29] Alternatively gels from other metal oxide such as vanadia (V_2O_5) can be used in which electronic conduction is believed to occur via valence hopping.[30]

6.4   CATALYSTS

Due to its porous nature silica has been extensively explored as a catalytic material. For this purpose advantage has been taken of two of its principal properties: silica contains a large number of surface hydroxyl groups that can act as acid catalysts, and the surface of silica can be easily loaded with catalytically active sites through impregnation with various metals.

As was already described in Chapter 3, sol-gel derived materials can be prepared in the presence of organic molecules leading to the imprinting of the template structure in the inorganic material. The imprinting of organic molecules that mimic the transition state of an organic reaction has led to highly selective catalytically active silicates. However, imprinting technology always is a trade-off between the accuracy of the imprint and the accessibility of the active site, and for this reason is not pursued extensively at this time.

The use of supramolecular assemblies as templates for the construction of well-defined structured mesoporous materials, however, has created a new direction in the field of catalysis. Since the discovery of the M41S family it was immediately realised that this type of materials could offer an interesting alternative to the existing zeolite structures. As was mentioned in Chapter 3, a large variety of pore structures have been imprinted in sol-gel derived silicates. When compared to the traditional zeolites, the mesoporous materials have the advantage that their pore size can be accurately controlled, ranging from 15 to 100 Å, and that their pore surfaces can be modified after synthesis. The incorporation of catalytic centres into these structures has been achieved in three different ways (a) by doping the sol-gel mixture with the appropriate metal precursor (b) by post-synthesis deposition of a metal on the silicate structure, and (c) by anchoring metal complexes via ligands that are covalently bound to the silica matrix (figure 6.4).

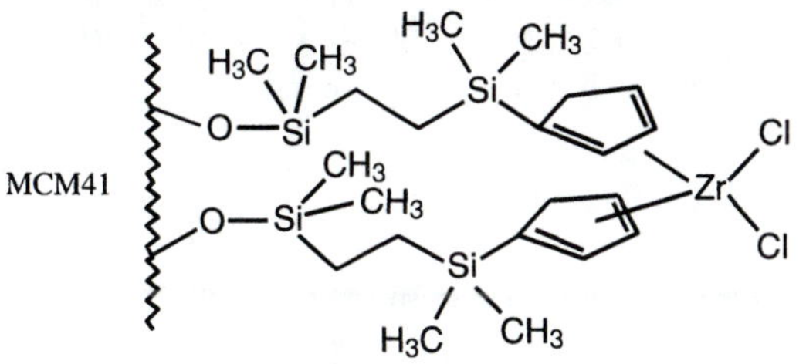

FIGURE 6.4   A METAL COMPLEX COVALENTLY BOUND TO THE SILICA MATRIX OF MCM 41.

The direct incorporation of metal ions into the silica framework by doping is probably the most elegant method. However, it suffers from the drawback that only low amounts of dopants can be introduced. For aluminium ions this is about 9%; for other metal ions the limit generally lies at 1–2%. These materials have been investigated as catalysts in many reactions,[31,32] of which a few illustrative examples will be discussed.

A titanium-doped silicate TiMCM-41 was prepared by the group of Corma.[33] This material showed a higher catalytic activity in the epoxidation of norbornene than two zeolites (Ti-β zeolite and Ti-ZSM-5). For the more sterically demanding substrate 2,6-di-*tert*-butyl phenol it was found that the Ti-HMS catalyst prepared by the group of Pinnavaia[34] had even better activity. These results were interpreted in terms of the lower diffusion limitations for the substrate due to the increasing pore sizes of the catalysts going from the zeolites to the TiMCM-41to the Ti-HMS material.

Tin-substituted mesoporous silica prepared by the group of Pinnavaia has been used for the ring opening polymerisation of lactides.[35] This catalyst gave a higher conversion than tin-doped silica or pure SnO_2. Compared to homogeneous catalysts Sn-HMS led to higher molecular weight and lower polydispersity of the polymer. The latter effect was attributed to the channel structure of the catalyst which can act as a "strait-jacket" for the polymer, preventing intermolecular transesterification reactions and "back-biting".

Abbenhuis and co-workers reported the immobilisation of a catalytically active titanium(IV) silsesquioxane in an MCM41 molecular sieve. This interesting approach leads to the generation of active, truly heterogeneous, and reliable catalysts for the epoxidation of alkenes.[36]

Mesoporous silicates can be loaded with a large variety of catalytic metal components using different methods of which incipient wet impregnation is the most common one.[37]

Other methods are exemplified by the preparation of Pt-MCM-41 by ion exchange with $[Pt(NH_3)_4Cl_2]$ followed by calcination[38] and the incorporation of titanium oxide by treatment of the silicate with a solution of $[(C_5H_5)_2TiCl_2]$ after which the retained titanocene complex was oxidised to TiO_2.[39] Wet methods, however, have the disadvantage that frequently unwanted cluster growth occurs during the deposition of the metal centers. This can be avoided by the use of vapour grafting using volatile organometallic precursors. Pd-TMS11 was prepared by vacuum sublimation of $[(C_5H_5)_2Pd(\eta^3\text{-}C_3H_5)]$ through the hexagonally packed mesopores.[40] This material was used to catalyse the Heck reaction and proved to be superior to other heterogeneous and many homogeneous Heck catalysts.

The immobilisation of catalytically active metal complexes in mesoporous materials generally involves the treatment of the silicate with a reactive functionalised alkyl trialkoxysilane which is then subsequently used as an anchoring point for the catalyst. As anchoring points 3-chloropropyltrialkoxysilanes,[41] 3-glycidyloxypropyl-trialkoxysilanes,[42] and 3-aminopropyltrialkoxysilanes have been used predominantly. In this way ruthenium porphyrin was attached to MCM-41 whose surface had been treated with 3-aminopropyltriethoxysilane.[43] When this catalyst was used for the oxidation of stilbene it was found to give predominatly the *trans*-isomer whereas the free porphyrin gives a 1:1 mixture of *cis*- and *trans*-stilbene oxide. This indicates once more the effect of the steric restrictions imposed upon the substrate by the mesoporous matrix. Moreover, the turn-over numbers for the immobilised catalyst were 20–40 times higher than that of the free porphyrin.

6.5   COATINGS

The prospects for the application of sol-gel technology are probably most pronounced in the field of coatings. The sol-gel approach offers a number advantages inherent to the technique, including the possibility to coat large, curved substrates using simple deposition techniques and the ability to obtain homogenous coatings, as well as coatings with tailored inhomogeneity. Sol-gel methods can be conveniently used to produce multilayer coatings. Furthermore, they allow the preparation of composites which cannot be obtained by other methods, such as organic-inorganic hybrid materials.

Sol-gel layers have been used to achieve conductive coatings, passivation coatings and porous coatings, optical-control- and antireflection-coatings, adhesion-promotion- and adhesion-resistance-coatings, biocompatible coatings,[44] antisoiling coatings and several others.[45,46] Here we will discuss some characteristics specific for sol-gel coatings.

Thin sol-gel layers can be applied to substrates using both spin coating and dip coating.[4,47] These two processes lead to distinctly different films due to the difference in the way the gel-layers are deposited. In the spin coating process the film is deposited and dried in a few seconds, whereas in the case of dip coatings the film is applied with a rate of typically a few centimetres per minute. Consequently dip coating leads to a better alignment of the polymer molecules, resulting in a denser, less rough film than in the case of spin coating. Both methods however have in common that there is an inverse relation between the thickness of the film and its density, *i.e.* thin films are denser than thick films. Furthermore, it is evident that the composition of the applied sol is important in determining the characteristics of the coating. Some rules of thumb are:

– Alkoxides of lower alkyls produce denser films with higher oxygen content.
– As in bulk materials, the pore size and surface area strongly depend on the solvent.
– Hydrolytic condensation in dilute solution leads to smaller particles and hence to a finer texture and a higher porosity.
– Higher water contents during hydrolysis lead to denser films.
– Aggregation of particles prior to deposition leads to highly porous films.

Spin coating can provide highly uniform coatings, however it is not a suitable method to deposit very thick films and is not convenient for the coating of large areas of unsymmetrical substrates. On the other hand, dip coating is a technique that is very well suited for such tasks, although this method frequently leads to a non-uniformity of the coating near the edges of the substrate.

6.6   MEMBRANES

The porosity required for filtration and separation on a sub-micrometre scale generally requires the use of membranes. Generally the mode of separation is size exclusion, but there are also membranes that actively separate on electrostatics, hydrophobicity or complexation. Most membrane materials are based on polymers or plastics due to their ease of fabrication. Ceramic membranes, however, have distinct advantages over organic materials, i.e. they do not decompose or swell like plastics and they are harder and more resistant to abrasion. There have been two principal ways of preparing porous ceramic membranes: by selective leaching of one of the components of a phase-separated glass, or by sintering of a glass compact leading to neck formation between the particles.[48] The sol-gel approach has the advantage that porous materials are inherent to this technique, that thin layers are easily prepared and that a large variety of pore sizes and surface chemistries are available. Silica-based top layers on γ-Al_2O_3 membranes are currently exploited for the separation of gases. Good results have been obtained for the separation of hydrogen and propene.[49] An extension of this technology is the fabrication of columns for size exclusion chromatography by the hydrolysis and condensation of TMOS in the presence of polyethylene oxide. The resulting materials are promising since they allow higher elution rates than commercial columns without loss of performance.[50]

REFERENCES

1.   D. Levy, L. Esquivas, *Adv. Mater,* **7**, 120 (1995).

2.   S. Sakka, *J. Sol-Gel Sci. Technol.*, **3**, 69 (1994).

3.   M. Toki, S. Miyashita, T. Takeuchi, S. Kanbe, A. Kochi, *J. Non-Cryst. Solids,* **100**, 479 (1988).

4.   C.J. Brinker, A.J. Hurd, P.R. Schunk, G.C. Frye, C.S. Ashley, *J. Non-Cryst. Solids,* **147&148**, 424 (1992).

5.   Y. Harruv, S.E. Webber, *Chem. Mater,* **3**, 501 (1991).

6.   M. Tarasevich, *Ceram. Bull.,* **63**, 500 (1983).

7.   J. Zarzycki, *Heterogen. Chem. Rev.,* 1 (1994).

8.   C. Bagnall, J. Zarzycki, in *Sol-Gel Optics-I,* edited by J.D. Mackenzie and D.R. Ulrich, *SPIE Proc.,* 1738 (1990).

9.   M. Canva, P. Georges, G. LeSaux, A. Brun, D. Larrue, J. Zarzycki, *J. Non-Cryst. Solids,* **135**, 182 (1992).

10.   S. Sakka, K. Kamiya, *J. Non. Cryst. Solids,* **48**, 31 (1982).

11.   M. Sekine, S. Katayama, *J. Non-Cryst. Solids,* **134**, 199 (1991).

12.   K. Tiefenthalerr, V. Briguet, E. Buser, M. Horisberger, W. Lukoz, *Proc. Photo-Opt. Instrum. Eng.,* **401**, 165 (1983).

13.   N. Tohge, A. Matsuda, T. Minami, Y. Matsuno, S. Katayama, Y. Ikeda, *J. Non-Cryst. Solids,* **100**, 501 (1988).

14.   R.L. Roncone, L.A.Weller-Brophy, L. Weisenbach, B.J.J. Zellinski, *J. Non- Cryst. Solids,* **128**, 111 (1991).

15.   D. Levy, D. Avnir, *J. Photochem. Photobiol. A.: Chem.,* **57**, 41 (1991).

16.   T. Tani, H. Namikawa, K. Arai, *J. Appl. Phys.,* **58**, 3559 (1985).

17.   D, Levy, S. Einhorn, D. Avnir, *J. Non-Cryst. Solids,* **113**, 137 (1989).

18.   P.N. Prasad, in *Proc. 4th Int. Conf. On Ultrastructure Processing of Ceramics, Glass and Composites,* edited by D.R. Uhlmann, and D. Ulrich. Wiley-Interscience: N.Y., 1990.

19.   D.R. Ulrich, *J. Non-Cryst. Solids,* **121**, 465 (1990).

20.   D. Avnir, D. Levy, R. Reisfeld, *J. Phys. Chem.,* **88**, 5956 (1984).

21.   R. Gvishi, R. Reisfeld, *J. Non-Cryst. Solids,* **128**, 69 (1991).

22.   G. Philipp, H. Schmidt, *J. Non-Cryst. Solids,* **63**, 283 (1984).

23.   R. Zusman, C. Rottman, M. Ottolenghi, D. Avnir, *J. Non-Cryst. Solids,* **122**, 107 (1990).

24.   K. Kimura, T. Sunagawa, M. Yokoyama, *Chem. Commun,* 745 (1996).

25.   T.V. Rattan, G.E. Bandini, A.W. Palmer, A.C.C. Tseung, *Sensors & Actuators A.,* **5**, 483 (1991).

26.   N.A.J.M. Sommerdijk, A. Poppe, C. Gibson, J.D. Wright, *J. Mater. Chem.,* **8**, 565 (1998).

27.   B.C. Dave, B. Dunn, J.S. Valentine, J.I. Zink, *Anal. Chem.,* **66**, 1120 (1994).

28. S.A. Yamanaka, F. Nishida, L.M. Ellerby, C.R. Nishida, B. Dunn, J.S. Valentine, J.I. Zink, *Chem. Mater*, **4**, 495 (1992).

29. P. Auderbert, C. Demaille, C. Sanchez, *Chem. Mater.*, **5**, 911 (1993).

30. V. Glezer, O. Lev, *J. Am. Chem. Soc.*, **115**, 2533 (1993).

31. J.Y. Ying, C.P. Mehnert, M.S. Wong, *Angew. Chem. Int. Ed. Engl.*, **38**, 56 (1999).

32. A. Corma, *Chem. Rev.*, **97**, 2373 (1997).

33. A. Corma, M.T. Navarro, J. Perez Pariente, *Chem. Soc Rev.*, 147 (1994).

34. P.T. Tanev, M. Chibwe, T.J. Pinnavaia, *Nature*, **368**, 321 (1994).

35. T.M. Abdel-Fattah, T.J. Pinnavaia, *Chem. Commun.*, 665 (1996).

36. S. Krijnen, H.C.L. Abbenhuis, R.W.J.M. Hansen, J.H.C. van Hooff, R.A. van Santen, *Angew. Chem. Int. Ed. Engl.*, **37**, 356 (1998).

37. A. Corma, A. Martinez, V. Martinez-Soria, J.B. Monton, *J. Catal.*, **153**, 25 (1995).

38. K.J. Del Rossi, G.H. Hatzikos, A. Huss Jr, *US-A 5256277*, (1993) [*Chem. Abstr.*, **119**, 31332 (1993)].

39. T. Maschmeyer, R. Frey, G. Sankar, J.M. Thomas, *Nature*, **378**, 159 (1995).

40. C.P. Mehnert, J.Y. Ying, *Chem. Commun.*, 2215 (1997).

41. P. Sutra, D. Brunel, *Chem. Commun*, 2485 (1996).

42. Y.V.S. Rao, D.E. De Vos, T. Bein, P.A. Jacobs, *Chem. Commun*, 355 (1997).

43. C.-J. Lui, S.-G. Li, W.-Q. Pang, C.-M. Che, *Chem. Commun*, 65 (1997).

44. M. Gerritsen, A. Kros, J.A. Lutterman, R.J.M. Nolte, J.A. Jansen, *Biomaterials* (In the press).

45. D.R. Uhlmann, T. Suratwala, K. Davidson, J.M. Boulton, G. Teowee, *J. Non-Cryst. Solids*, **218**, 113 (1997).

46. D.R. Uhlmann, G.P. Rajendran, *SPIE Proc.*, **1328**, 270 (1990).

47. B.E. Yoldas, *J. Sol-Gel Sci. Technol.*, **1**, 65 (1993).

48. L.C. Klein, Design of microstructures in sol-gel silicates in *Design of New Materials Ceramics*, edited by A. Clearfield, D.L. Cocke. Plenum Press: New York, 1986.

49. R.J.R. Uhlhorn, K. Keizer, A.J. Burggraaf, *J. Membrane Sci.*, **66**, 271 (1992).

50. K. Nakanishi, H. Minakuchi, N. Soga, *J. Sol-Gel Sci. Technol.*, **8**, 547 (1997).

CHAPTER 7

APPLICATIONS OF METAL OXIDE SOL-GELS

7.1 INTRODUCTION

As indicated in Chapter 1, many of the earliest applications of the sol-gel method involved metal oxide materials. Significantly, these applications fell largely into two categories: decorative coatings and constructional materials. Today, while our improved understanding of the sol-gel process has been applied to the development of new generations of ceramic materials, new applications of metal-oxide-based functional materials with a higher level of sophistication have greatly widened the scope of sol-gel technology. New catalysts made by the sol-gel method attract the attention of chemists and chemical engineers; new electronic materials accessible via the sol-gel route find wide application in electronic devices and displays; new metal oxide semiconducting sensor materials offer improved sensitivity and selectivity, providing a link between the chemical environment and the computer. In contrast to the historical applications of the method, these new applications demand a detailed understanding of both the principles of the applications area and the principles of the sol-gel method if successful synergies are to develop. In this chapter we illustrate this process with reference to ceramic materials, catalysts, electronic materials and sensor materials.

7.2 CERAMICS

A ceramic is generally defined as a material which is first shaped (e.g. as a bulk object or as a coating) and then hardened by heating. The shaped object before heat treatment is known as a "green body". According to this view, many functional materials such as catalysts and electronic or sensor materials could be classed as ceramics. However, for the purposes of this text such materials will be treated separately, according to their functional value, and ceramics will be classed as materials defined in the above way but developed for their mechanical, chemical and thermal properties. Typical examples are light, strong, thermally-resistant materials for aviation, defence, space and sports applications, chemically resistant coatings for the chemical industry, and hard low-friction thermally resistant coatings for engine parts.

As explained in chapter 4, the formation of small uniform particles with minimum void regions between the particles is a critical factor in the production of strong ceramics. There are a number of other reasons[1] why small particles are preferred in ceramic "green bodies". Firstly, the sintering and densification temperatures are lower if small particles are used. For example, sol-gel methods can be used to prepare ThO_2-UO_2 spheres at densification temperatures of 1150°C, which compares with the value of 1700°C typically required if non-sol-gel powders are used.[1] Secondly, densification is often observed without associated increases in grain size which would lead to mechanical weakness at the inter-grain voids. For example, nanocrystalline titanium nitride densifies between 1100–1300°C, whereas grain size increase is negligible until above this temperature. Both of these observations may be due to the high volume fraction of surface regions of nanoparticles, whose ready rearrangement leads to sintering and densification at temperatures where the crystalline ordered cores of the nanoparticles are retained. An alternative view is that for uniform spherical particles each particle is in a close-packed array with 12 nearest neighbours and on sintering forms 12 "necks" joining it to the neighbours, with minimal probability of the uneven pore growth which would be expected for an array of non-uniform or larger particles. Finally, densification with minimal increase in grain size reduces light scattering at grain interfaces and leads to improved optical properties.

The chemical problem in generating materials with small particles for ceramic fabrication is that there are two conflicting requirements: on the one hand small particles are required, which are to be prevented from aggregating; on the other hand, high green-body solid densities are preferred, and methods which prevent aggregation by generating a protecting ionic or polymer layer around the particle will tend to give a low solid density. An effective solution to this dilemma has been found[2] by surface modification of the nanoparticles with small molecules such as carboxylic acids (e.g. methacrylic acid stabilises 2nm zirconia particles) or amines (which stabilise gold, silver, palladium and copper colloids). The green-body density of the concentrated colloidal suspensions of such particles can be typically 60% solid content by volume, and the resulting materials can be moulded, pressed or extruded into a variety of shapes before firing to densify. For example, nanocrystalline boehmite stabilised with propionic acid can be concentrated to a plastic material containing 40–50 volume % of the solid material, and extruded and fired to form alumina tubes.

Even with careful control of particle size, some additional toughening of ceramic materials is often desirable for the advanced applications mentioned above. This is commonly achieved by forming ceramic composites, which may be of two types. In the first type, dispersed particles block the propagation of cracks by diverting or arresting their movement. Here there is scope for varying the shape of the dispersed particles in a controlled way, and recent developments in the use of adsorbed molecules to control crystal growth and morphology in organic crystals, and in the understanding of biomineralisation processes, may well find application in providing fine control over the dispersed particles. The second type is that of fibre-reinforced composites. Here fibres (which may be made from sol-gel-derived preforms or extruded ceramic precursors, or by chemical means – e.g. SiC) are entrapped in the ceramic green body before densification. Load transfer from the ceramic matrix to the fibres can substantially improve breaking stress. Although the use of the sol-gel method here is potentially valuable because the sol-gel can be cast around the fibres in a mould to give desired shapes, the problem of shrinkage on densification is serious. The fibres effectively resist shrinkage and can lead to cracking of the surrounding matrix.

7.3   CATALYSTS

Although, as pointed out in chapter 1, the use of sol-gel methods in catalyst preparation dates back to the work of W.A. Patrick in the 1920's, the significance of catalysts in the world economy has grown enormously since then. More than 90% of the processes carried out in the chemical industry worldwide involve use of catalysts, and catalysts are involved in the generation of about 20% of gross national product in industrialised countries. In heterogeneous catalysis, the aim is to maximise the catalytic activity and reaction rate for reactions which occur at known active sites on the solid surface. Key parameters which influence these quantities and which are in principle controllable via sol-gel processing of the catalyst materials are as follows:[2]

i)   High specific surface area
ii)   Controlled pore-size distribution
iii) Stable pore structure under preparation and reaction conditions
iv)   Active material distributed on pore surfaces, not in the bulk
v)   Active material homogeneously and effectively distributed over surface
vi)   High purity of active material
vii) Easy control of composition for multi-component catalytic species
viii) Effective control of crystalline or amorphous structure as desired
ix)   Mechanical properties adapted to desired operating environment
x)   Catalyst resistant to chemical or physical blocking of active sites.

From the material already covered in this book, it will be immediately obvious how sol-gel processing can achieve many of these characteristics. Thus for example it should be clear that (i), (ii), (iii), (vi), (vii) and (viii) can all be achieved to some extent using the sol-gel method. However, in addition the sol-gel method opens up several entirely new possibilities for catalyst design. For example, as mentioned in chapter 6 the mild low-temperature conditions for preparation of silicate sol-gels permit the entrapment of enzymes and even whole biological cells, opening new prospects for bio-catalysed systems.[4] Furthermore, in a logical development of biological ideas, it is possible to entrap organic species within a sol-gel matrix and then remove the species, leaving behind a "footprint" catalysis site. In this way, materials have been prepared which contain imprinted transition-state analogues which show activities comparable to those of well-known enzymes.[5] Yet another novel use of sol-gel material properties is found in the use of microporous (diameter < 2nm) amorphous sol-gel glasses containing 1% platinum to provide poison resistant hydrogenation catalysts. The very small pore sizes allow only hydrogen activation, and no ingress of poisons or generation of entrapped side-products which might physically block the active sites, since reaction of the activated hydrogen occurs at the outer surface of the membrane.[6]

A fully comprehensive review of applications of sol-gel methods in catalysis is clearly impossible in a text of this size. Instead we summarise some of the key advantages and problems of preparing and using such materials as catalysts.

From the late 1960s aerogels became accessible as catalysts[7] due to the realisation by Nicolaon and Teichner[8] that the preparation process could be speeded up considerably by the use of alcoholic solutions of alkoxides, acetates and acetylacetonates with water only present as a reactant rather than as a solvent. These precursors also removed the need to wash out the counter-anions inevitably present with conventional inorganic salt precursors. The small amount of water present minimised the time required for solvent exchange prior to supercritical drying. (This solvent exchange is necessary because the critical point for water is at very high temperature and pressure (374°C and 219 atm.) and under these conditions recrystallised materials with decreased surface area and porosity are obtained). The resulting aerogel catalysts showed much better resistance to thermal treatment than the corresponding xerogels; for example a silica aerogel heated to 500°C is reported to still have a surface area of 800 m^2/g.[7] More recently, fast sol-gel methods have been developed which permit gel formation within a few minutes. For example, nickel/alumina aerogels have been prepared from gels made in a few minutes by mixing an ethanol solution of partially hydrolysed aluminium *sec*-butoxide with an ethanolic nickel acetate solution and hydrolysing with a water/ethanol mixture.[9] Such mixed oxide sol-gel catalysts have been found to have several advantages over single oxide materials: they have enhanced thermal stability, more stable surface area and pore volume, and improved catalytic activity. An example is the formation of mixed WO_3/ZrO_2 samples containing 6, 18 and 36 weight % of WO_3 from the mixed iso-propoxide precursors.[10] Subtle variations on the sol-gel procedure can be achieved by combining the sol-gel method with impregnation methods, giving a range of catalyst structures and activities. Thus the Ru-Sn-Al_2O_3 system has been prepared (a) by co-hydrolysis of mixed solutions of the precursors in organic solvents, (b) by treatment of the Ru-Al_2O_3 product prepared in the same way with a solution of the tin precursor, (c) by treatment of the Sn-Al_2O_3 product with a solution of the Ru precursor, or (d and e) by consecutive treatment of sol-gel-prepared alumina with Ru and Sn precursors in two ways, with Ru or Sn treatment first respectively. The five catalysts showed different surface compositions and different selectivity as catalysts for hydrogenation of dimethyl-terephthalate.[11]

Aerogel catalysts incorporating dispersed metals promote activation of hydrogen and subsequent "spillover" of this activated hydrogen to generate new active sites on the oxide surface, and this is a very important mechanism in catalytic hydrogenation reactions. It has been argued that in some cases hypercritical conditions can alter the microporosity of silica particles which contain entrapped metallic catalyst particles, leading to loss of catalytic activity.[12] In such cases low density xerogels may provide more effective catalysts, and recently efforts have been made to develop improved fabrication methods for xerogel catalysts, both by reducing the surface tension of the liquid phase and by increasing the mechanical strength of the wet gel to resist the capillary forces developed on drying.[12–14] These approaches should have cost advantages over methods involving supercritical drying, as well as offering possible different pore size distributions. However, unless the above-mentioned improved methods are adopted, xerogels generally have pore volumes an order of magnitude smaller than those of aerogels.[7]

Some of the properties which make sol-gel materials so effective as catalysts also provide potential problems in the development of practical catalytic reactors. The high porosity also leads to a low material density, good thermal insulation properties and a tendency to develop large pressure drops across the catalyst due to the small-pore structure. The aerogel product from supercritical drying is also often a fine powder or loose fragile lump structure. Some of these problems can be overcome by supporting the aerogel on or in a matrix with higher thermal conductivity. Many of the problems can also be overcome by the use of fluidised bed reactors, although high gas velocities are needed for fluidisation[7] and this can cause problems in view of the low density of the material.

7.4 ELECTRONIC MATERIALS

Sol-gel methods offer significant general advantages in fabrication of electronic materials:

i)   High purity is attainable via purified pre-cursors
ii) Uniform materials of controlled complex composition can be obtained by the use of suitable homogeneous precursor mixtures
iii) Relatively low processing temperatures may be used, minimising reaction with substrates
iv) Films may be prepared by spin coating without the need to involve vacuum chambers or the high energy costs of vacuum evaporation, and large area substrates (e.g. semiconductor wafers) can be rapidly coated.

However, a number of problems occur, notably the difficulty of obtaining thicker films (typically 10–20μm) for micro-machined components, the prevention of substrate structure adversely affecting the nucleation of the desired solid phase from the sol-gel film, and the possibility that reaction and/or interdiffusion may occur on heat treatment to nucleate the desired phase despite the lower temperatures generally required than for alternative processes.

7.4.1 Ferroelectric Materials

The above advantages and limitations are well illustrated by the formation of ferroelectric lead zirconium titanate (PZT) $PbZr_xTi_{(1-x)}O_3$ films. This material shows switching characteristics, making it suitable as an electronic memory material, as well as piezoelectric and pyroelectric properties offering application in actuator and sensor devices, and a high dielectric constant which facilitates fabrication of miniaturised capacitor components. Using precursors such as lead acetate trihydrate with *n*- or *iso*-propoxides of zirconium and titanium in a suitable solvent (e.g. 2-methoxyethanol or acetic acid), spun films typically 0.3µm thick can be prepared, and the evolution of the desired perovskite phase on rapid thermal annealling has been studied.[15] Even if more viscous precursor mixes are used, it is still difficult to obtain films thicker than 1µm. Initial attempts to prepare thicker films by floating on water a layer of a precursor solution in a solvent which is immiscible with water have produced films 10–20µm thick, but substantial shrinkage was observed leading to creasing of the film before it could be laid down on a substrate.[16] A promising alternative has been shown to be the use of powders prepared from dried mixtures of the above-mentioned precursors, which are then dissolved in mixtures of 1,3-propanediol, triethanolamine and water at 90°C to give sols which can be spread by a doctor-blade technique. Using reproducible multi-layer deposition, this method has been shown capable of producing PZT films over 5µm thick.[17] The problem of ensuring absence of substrate effects on crystallisation has been approached by depositing thin layers of sol-gel based materials which are known to crystallise readily into the desired perovskite phase, before spin-coating with the desired PZT material. Such layers perform the dual function of acting as epitaxial crystallisation templates and diffusion barriers.[18]

There is growing interest in related ferroelectric materials for use in electronic devices, due to problems arising from the volatility of PbO even at processing temperatures of 550–650°C, leading to modified crystal structure and oxygen vacancies. These related materials include, for example, $LiNbO_3$, $YMnO_3$, $SrBi_2Nb_2O_9$, $SrBi_2Ta_2O_9$, $BaTiO_3$ and $Ba_{0.7}Sr_{0.3}TiO_3$, which have all been prepared using sol-gel methods.[19] One further example of the use of the sol-gel method to prepare complex oxide mixtures is its application to the preparation of Bi-Sr-Ca-Cu-O high-temperature superconductors from acetate precursors.[20]

7.4.2 Electrochromic Materials

Another class of electroactive materials with widespread commercial application which can be prepared by the sol-gel process is electrochromic materials[21,22] including WO_3, V_2O_5, Nb_2O_5, TiO_2, CeO_2, Fe_2O_3 and mixtures such as CeO_2/TiO_2, CeO_2/SnO_2 and WO_3/TiO_2. In these materials, electrochemical insertion of H^+ or Li^+ leads to formation of a coloured partially reduced mixed-valence material. Typically the electrochromic layer is deposited on ITO-coated glass and combined with a counter-electrode layer (providing charge-storage and also deposited on conductive glass) via an electrolyte layer (which may be either a solid or liquid electrolyte). Although such materials were first prepared by conventional methods, the sol-gel method offers significant advantages in preparing high-quality coatings of controlled composition and homogeneity with high surface area and high porosity, which can bind large amounts of ions and provide rapid ionic diffusion. Amorphous hydrous gels, in addition to offering fast ionic diffusion, can more easily accommodate volume changes associated with the redox reactions responsible for electrochromism, adhere well to ITO conductive coatings and can be deposited under ambient temperature and pressure conditions. (The latter advantage is not a very large one in practice, however, since the most costly process is the ITO coating.) For example, a sol prepared from a tungsten peroxyester gave a WO_3 film on heat treatment at a temperature as low as 100°C, with transmission electrically switchable from 15% to 85% within a few seconds. Such films are available in a range of colours, and find application in anti-dazzle rear-view mirrors, electrochromic sunroofs and sunglasses, and electrochromic windows for climate control, and the sol-gel method is still under development in this area.

7.4.3 Solid Electrolytes

Solid electrolytes, mentioned above in the construction of electrochromic devices, have also been successfully prepared using sol-gel methods. Typical examples include the proton conducting zirconium and titanium phosphates, antimonic acid $HSbO_3nH_2O$ and $V_2O_5.nH_2O$, the sodium ion conducting systems $Na_{1+x}Zr_2Si_x$ $P_{3-x}O_{12}$ ("Nasicon"), $Na_5YSi_4O_{12}$, $Na_5GdSi_4O_{12}$ and $Na_4Zr_2Si_3O_{12}$, the lithium ion conducting Li-salt-doped silica gels, Li aluminosilicates, gallosilicates and borosilicates, and compounds such as $Li_{1.3}Al_{0.3}Ti_{1.7}(PO_4)_3$ and novel oxide ion conductors such as $La_{0.9}Sr_{0.1}Ga_{0.8}Mg_{0.2}O_{2.85}$. The advantages of using sol-gel methods for such materials include formation of amorphous glassy materials of defined and controllable composition at low temperatures, isotropic conductivity, good ionic conductivity combined with low electronic conductivity, and ease of fabrication.[22]

7.4.4 Other Electronic Materials

The electronics industry also has requirements for materials which are of less dramatic electronic characteristics but no less important for the successful implementation of new devices. Typical of these are transparent electrically conducting coatings and substrates having very high thermal conductivity combined with good electrical insulation. The use of indium oxide and tin-doped indium oxide (ITO) as a transparent electrically conductive film in devices such as liquid crystal displays is well-known. However there is now a need for such films to be prepared using low-temperature methods for application to polymers and other organic films. Although sol-gel methods, starting from *sec*-butoxides of tin and indium, can be used to prepare transparent conductive films, heat treatment is usually required for crystallisation and removal of residual organic and hydroxyl groups. Recent reports suggest that the latter processes may be achieved without the need for high temperatures, by the use of UV-laser irradiation of relatively low power.[23] Sol-gel methods have also been successfully been investigated for the preparation of ultrafine alumina powders[24] as possible precursors for reaction to yield aluminium nitride, which has extremely high thermal conductivity (90–190W.m^{-1}K^{-1}, approaching the values for metals (e.g. Cu 400 W.m^{-1}K^{-1}) while retaining good electrical insulation (10^{11}–$10^{14}\Omega$.cm), and a coefficient of thermal expansion (4.3–4.5 × 10^{-6}K^{-1}) close to that of silicon. The availability of this material at reasonable cost would provide a significant advantage for the preparation of new high-performance computer chips, where one of the ultimate performance-limiting factors is the rate at which heat can be dissipated from the active switching elements. These and other electronic applications of sol-gel materials are excellent examples of areas where the advantages of the method are most likely to overcome disadvantages such as high precursor cost and long processing times, because of the high product value and small amounts of materials per device.

7.5 FLAMMABLE GAS SENSORS

Semiconducting metal oxides are an important class of gas sensing materials. Many metal oxides are semiconductors because of the variable valency of the metal leading to non-stoichiometry and cation or anion vacancies. A widely studied example is tin oxide, which is generally an n-type semiconductor. Formally, some of the tin sites have a tendency to exist as Sn(II) or even Sn(0), leading to a deficiency of oxygen to maintain charge neutrality. The anion vacancies formally adopt a negative charge as the lower oxidation state tin species act as electron donors, and it is these negative charges which provide the majority charge carriers in the semiconducting material. However, semiconducting oxides also possess two other important properties: gases can adsorb on the solid surface, and the surface can catalyse reactions between different adsorbed gases. Notably, in the presence of the weak electron acceptor oxygen, weak charge-transfer chemisorption takes place, resulting in some of the mobile negative conduction electrons in n-type semiconductors such as tin oxide being trapped on surface oxide species. The semiconductivity in the presence of oxygen is therefore smaller than that for a sample with a clean surface in a vacuum. If a flammable gas approaches the surface in the presence of oxygen, surface catalysed reactions occur between the flammable gas molecules and surface oxide species, leading to removal of the surface oxygen species and return of the trapped charge carriers to the conduction state. The semiconductivity thus increases, to an extent proportional to the concentration of the flammable gas and the rate of reaction.

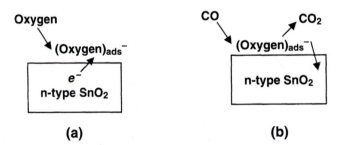

(a)   **(b)**

FIGURE 7.1 (A) OXYGEN ADSORPTION DECREASES CONDUCTIVITY BY SURFACE TRAPPING OF CONDUCTION ELECTRONS. (B) CONDUCTIVITY INCREASES AS (OXYGEN)$^-$ REACTS WITH FLAMMABLE GAS AND TRAPPED ELECTRONS ARE RETURNED TO THE CONDUCTION BAND.

This simple mechanism, illustrated in figure 7.1, explains both how the materials function as gas sensors and why the sensing process is dependent on the nature of the flammable gas, the temperature of the sensor surface and the presence of additional catalysts. Thus, gases such as hydrogen and carbon monoxide, which react easily with oxygen, are detected with high sensitivity at low sensor temperatures, whereas gases such as methane and other hydrocarbons (whose combustion reactions have high activation energies due to the number of chemical bonds which must be broken and formed) require a higher sensor temperature. For all flammable gases, the sensitivity ultimately falls off at very high temperature due to thermal desorption of the reacting species from the surface. The presence of well-known combustion catalysts such as Pd and Pt on the metal oxide surface also increases the sensitivity at low temperatures. (The sensitivity is commonly defined as the change in conductivity on exposure to a given concentration of flammable gas, expressed as a percentage of the initial conductivity in clean air at the same temperature.)

Although very extensive studies on a wide range of metal oxide sensor materials[25,26] have led to many more detailed and complex mechanisms, including consideration of both surface and bulk processes and grain boundary effects as well as those of the above simplified model, in all cases surface area, surface composition and overall stoichiometry of the oxide material are important parameters determining gas sensor performance. Hydrolytic processing of metal oxides, as discussed in chapter 4, provides opportunities for control of these parameters which are not available where simpler high-temperature oxidation processes are used. Thus, low temperature hydrolysis and condensation can lead to very small particle sizes (*nanocrystals*) and hence very large surface areas maximising surface reactions and hence sensitivity, whereas the use of high temperature processes leads to larger particles due to sintering, regardless of the size of the initially-formed particles. Furthermore, from the above-mentioned definition of sensitivity, high sensitivity is favoured by a low initial semiconductivity, which in turn is favoured by samples whose chemical compositions are closer to ideal stoichiometry. Since non-stoichiometry is a state of higher entropy than the ideal stoichiometric state, it is favoured by high temperature. Hence, low-temperature processing leads to less non-stoichiometry and potentially higher sensitivity. Finally, low-temperature homogeneous solution processing of mixed oxide precursors can lead to samples with uniform controllable chemical doping of one oxide into another, with control of both bulk and surface composition. On heating, diffusion of dopants from bulk to surface can occur, with effects on both bulk and surface composition and on subsequent sintering processes. Samples prepared and used at low temperatures therefore offer many advantages as semiconductor sensors.

These principles will be illustrated with reference to nanocrystalline tin oxide sensor materials. Hydrolysis of sterically-hindered tin alkoxides (e.g. tetra-*t*-butoxy tin(IV)) in the presence of the chelating agent acetylacetone and polyethylene oxide led to high surface area nanocrystalline material.[27] Treatment with strong nitric acid removed acetylacetone and polyethylene oxide, increasing the accessible surface area. These samples were shown to have high sensitivity at low operating temperatures as sensors for flammable gases including solvents such as toluene, and to display sharper profiles of sensitivity as a function of temperature (and hence better selectivity) than currently available commercial tin oxide sensors.[28,29] Similar samples can also be prepared by hydrolysis of hydrated tin chloride, or by a sol-gel process starting from anhydrous tin chloride.[30] A detailed combined powder X-ray diffraction and EXAFS study has been reported[31] for such samples, both pure and doped with Cu^{2+} and Fe^{3+}, while heating between room temperature and 900°C. At low temperatures the structure and level of disorder in all samples was comparable to that in bulk tin oxide samples, with the dopants located in normal Sn^{4+} sites in the cassiterite lattice. On heating, sintering begins near 400°C for the pure samples, but later for the doped samples. EXAFS data suggest that at the higher temperatures the dopant ions move from their initial bulk sites to surface sites, where they inhibit the sintering process. CO sensing experiments on the same samples[32] revealed a temperature profile of sensitivity with two peaks, one at 200°C and the other around 400°C. For the pure tin oxide samples, sintering at 900°C increases the particle size much more than for the doped samples and reduces the size of the low-temperature sensitivity peak. However, the doped samples show much more significant loss of sensitivity in the low-temperature region, so the effect is not simply due to increased particle size but may involve such mechanisms as blocking of diffusion into the bulk when the dopant ions migrate to the surface, or dominance of the surface dopant ions in determining the reaction mechanism. These studies confirm the value of nanocrystalline samples in providing new gas sensing properties, but they also provide an important warning. If such materials are operated for long periods at temperatures of 400°C or above, interlinked sintering and dopant diffusion effects may permanently modify the sensing characteristics. Similar complications have been observed for palladium-salt doped tin oxide nanocrystals prepared by sol-gel methods, where the palladium salt led to smaller crystallite sizes following heating to 600°C and was shown to be incorporated into the sample both as bulk and surface Pd^{2+} ions and as metallic palladium. Although there have been relatively few reports of sensor materials prepared by these methods, it is clear that there is much scope for development of better sensors via this route as understanding of the mechanisms and sintering and diffusion processes increases.

REFERENCES

1.   D.L. Segal, *"Chemical synthesis of advanced ceramic materials"*, Cambridge University Press: Cambridge, 1989.
2.   H.K. Schmidt, "Chemical routes to nanostructured ceramics and composites" in *"Applications of organometallic chemistry in the preparation and processing of advanced materials"*, edited by J.F. Harrod and R.M. Laine, p. 47–67. Kluwer: Dordrecht, 1995.
3.   M.A. Cauqui and J.M. Rodríguez-Izquierdo, *J. Non-Cryst. Solids*, **147–8**, 724–738 (1992).
4.   D. Avnir, S. Braun, O. Lev and M. Ottolenghi, *Chem. Mater.*, **6**, 1605 (1994).
5.   W.F. Maier and J. Heilmann *Angew. Chem. Int. Ed. Engl.*, **33**, 471–3 (1994).
6.   W.F. Maier, F.M. Bohnen, J. Heilmann, S. Klein, H-C. Ko, M.F. Mark, S. Thorimbert, I-C. Tilgner and M. Wiedorn, in *Applications of Organometallic Chemistry in the Preparation and Processing of Advanced Materials*, edited by J.F. Harrod and R.M. Laine, p. 27–46. Kluwer: Dordrecht, 1995.
7.   G.M. Pajonk, *Applied Catalysis*, **72**, 217–266 (1991).
8.   G.A. Nicolaon and S.J. Teichner, *Bull. Soc. Chim. Fr.*, 1906 (1968).
9.   D.J. Suh, T-J. Park, J-H. Kim and K-L. Kim, *J. Non-Cryst. Solids*, **225**, 168–172 (1998).

10. M. Signoretto. M. Scarpa, F. Pinna, G. Strukul, P. Canton and A. Benedetti, *J. Non-Cryst. Solids*, **225**, 178–183 (1998).

11. M. Toba, S. Tanaka, S. Niwa, F. Mizukami, Z. Koppány and L. Guczi, *J. Sol-Gel Sci. Technol.*, **13**, 1037–1041 (1998).

12. A.J. Lecloux and J.P. Pirard, *J. Non-Cryst. Solids*, **225**, 146–152 (1998).

13. S. Haereid, M. Dahle, S. Lima and M.-A. Einarsrud, *J. Non-Cryst. Solids*, **186**, 96 (1995).

14. S. Haereid, E. Nilsen and M.-A. Einarsrud, *J. Non-Cryst. Solids*, **204**, 228 (1996).

15. I.M. Reaney, D.V. Taylor and K.G. Brooks, *J. Sol-Gel Sci. Technol.*, **13**, 813–820 (1998).

16. M. Yamane, *J. Sol-Gel Sci. Technol.*, **13**, 821–825 (1998).

17. P. Löbmann, S. Seifert, S. Merklein and D. Sporn, *J. Sol-Gel Sci. Technol.*, **13**, 827–831 (1998).

18. D.P. Birnie III, M.H. Jilavi, T. Krajewski and R. Nass, *J. Sol-Gel Sci. Technol.*, **13**, 855–859 (1998).

19. R.W. Schwartz, *Chem. Mater.*, **9**, 2325–2340 (1997).

20. Y. Masuda, R. Ogawa, Y. Kawate, T. Tateishi and N. Hara, *J. Mater. Res.*, 7, 292 (1992).

21. M.A. Aegerter, C.S. Avallaneda, A. Pawlicka and M. Atik, *J. Sol-Gel Sci. Technol.*, **8**, 689–696 (1997).

22. O. Lev, Z. Wu, S. Barathi, V. Glezer, A. Modestov, J. Gun, L. Rabinovich and S. Sampath, *Chem. Mater.*, **9**, 2354–2375 (1997).

23. H. Imai, A. Tominaga, H. Hirashima, M. Toki and M. Aizawa, *J. Sol-Gel Sci. Technol.*, **13**, 991–4 (1998).

24. E. Ponthieu, E. Payen and J. Grimblot, *J. Non-Cryst. Solids*, **147**, 598–605 (1992).

25. *Gas Sensors: Principles, Operation and Developments*, edited by G. Sberveglierli, Chapters 2–4. Kluwer: Dordrecht, 1992.

26. D.E. Williams and P.T. Moseley, *J. Mater. Chem.*, 1, 809–814 (1991).

27. C. Roger and M.J. Hampden Smith, *J. Mater. Chem.*, 2, 1111–1112 (1992).

28. A. Wilson, J.D. Wright, J.J. Murphy, M.A.M. Stroud and S.C. Thorpe, *Sensors and Actuators B*, **18–19**, 506–510 (1994).

29. A. Wilson and J.D. Wright, Improvements in Tin Oxide Based Gas Sensors, *UK Patent Application GB9319456*

30. M. Foster, J. Eberle, S. Straessler and P. Pfister, *New Materials and their Applications, IOP Conference Series No. 111*, 479–480 (1990).

31. S.R. Davis, A.V. Chadwick and J.D. Wright, *J. Phys. Chem. A*, **101**, 9901–8 (1997).

32. S.R. Davis, A.V. Chadwick and J.D. Wright, *J. Mater. Chem.*, **8**, 2065–2071 (1998).

CHAPTER 8

THE FUTURE

8.1 THE PLAYERS

The future prospects of sol-gel chemistry will concern at least two main groups of potential readers of this text. The first group is the new generation of scientists using this text as part of their training and considering whether to commit their scientific careers to this area. The second group consists of those in industry considering investment in this area of technology and seeking perspectives on future applications and markets. The decisions of both groups are interdependent. Future prospects for scientists in the field depend greatly on the economic prospects for development of successful commercial products which use sol-gel processes and materials to a sufficient extent to support a thriving future research activity. Future prospects for successful new applications of sol-gel technology depend on the availability of skilled researchers able to find solutions to outstanding obstacles to successful commercial applications and to develop innovative new high-performance materials which will in turn create new markets.

8.2 MARKET PROSPECTS

From the preceding chapters the range of materials advantages and possible applications offered by the sol-gel method can be seen to be very extensive. Low temperature solution processing with high purity, the facility of precise control of complex chemical composition and physical structure, control of pore structure and pore-wall chemistry, ready preparation of thin films and coatings or nanocrystalline powders, versatile inorganic/organic composite preparation and availability of good optical quality materials are all valuable attributes for development of new functional materials. Despite this there are few, if any, commercial applications of the sol-gel method with annual sales exceeding $30 million.[1] Furthermore, it will also be apparent from earlier chapters that there are several key areas where improved fundamental understanding or practical materials preparation methods are still needed before the above attributes can be fully exploited, even for currently anticipated applications such as thick ferroelectric films. Thus, the best short-term future prospects for sol-gel materials lie where the sol-gel method offers a clear advantage (e.g. for preparation of coatings and fine powders), or where the unique properties of sol-gel materials can be exploited (e.g. in materials which demand tailored porosity), or in totally new applications where there are no competing technologies. To imagine the latter, we need to consider the future prospects for research directions in sol-gel chemistry and processing. To develop new functional sol-gel materials, improved mechanistic understanding, a wider range of precursors and new processing methods are all desirable.

8.3 DEVELOPMENTS IN CHARACTERISATION METHODS

For a better mechanistic understanding, more widespread application of the newer powerful sol-gel characterisation methods is necessary. These include solid state NMR, X-ray techniques (including SAXS, diffraction and EXAFS) and pore volume/pore size distribution characterisation. The latter presents significant obstacles, since all of the available methods suffer some disadvantages.[2] Thus gas adsorption isotherm measurements give results which can depend on the gas used and which are complicated by the wide variety of surface character in a solid consisting of intersecting pores of varying size and shape, while thermoporometry and forced-mercury porometry lead to compressive stress and structural distortion and potential pore damage. These problems will if anything be more serious for more complex porous sol-gel materials prepared from mixed precursors, for example to give control of pore-wall hydrophobic character. Improved methods are needed, not only for determining pore sizes and distributions but also for determining the spatial distribution of pores within materials. For example, recently NMR imaging methods have been applied to give images of the distribution of pores of a given size within porous solids.[3] As improved measurements of pore size become available, better models for the control of pore size by variation of precursors and processing conditions should be developed, leading to improved materials in future. Controlled porosity, both in the initial gel and during stages of drying and densification, is a critical parameter in future developments of sol-gel materials.

8.4 COMPOSITE MATERIALS

As such methods are increasingly used, new and unexpected effects may become evident which open new avenues for control of sol-gel material properties. For example, solid state NMR studies of composites formed between metal oxide and silica sol-gels have shown in several cases that the metal oxide particles disrupt the silica structure on heating the sample, so that the number of Q^4 species actually decreases, while XRD studies of particle size changes in the same samples confirm that no growth of the metal oxide particles occurs, presumably because of their strong chemical interactions with the silica framework. Similar phenomena can also occur in zeolites, where particles synthesised with the zeolite may actually be larger than the normal pore size of the zeolite due to such disruptive interactions. Gel-guest interactions are also increasingly being studied in relation to the properties of composites containing entrapped organic molecules in silica or other oxide gels.[4] Entrapped molecules may adhere to pore walls by electrostatic, covalent or H-bonding interactions, their mobility and conformations may be restricted by the size and shape of the pores and their chemical and spectroscopic properties may be modified by the changed solvent structure and limited translational entropy within confined pores. These effects have important consequences for the design of chemical sensor materials consisting of sol-gel entrapped reagents,[5] since the associated thermodynamic changes can modify chemical selectivity as well as receptor availability. There is also evidence of significant interactions between polymers (e.g. polyethylene glycol) and silica gel. Anti-reflection coatings for laser optics based on sol-gel silica incorporating polyethylene glycol show improved resistance to laser irradiation damage, although there is so far no satisfactory model explaining the mechanism of damage limitation.[6] Clearly, a strong area for future development is the improved understanding of gel-guest interactions, and the consequent development of improved composite materials.

8.5   NEW PRECURSORS

Despite the considerable research activity in sol-gel chemistry, the range of precursors used in the majority of studies is still extremely small. By far the most common silicon gel precursor is TEOS, but even for this material there are new chemical aspects to be discovered, as was evidenced recently by the finding that aerial oxidation leads to a variety of impurities including acetaldehyde, acetic acid, silicon acetates and other products. The acetic acid and silicon acetate content has been found to be correlated with increased particle sizes, with as little as 0.1% acetate increasing the particle size of ammonia catalysed materials by a factor of 5, with consequent significant deterioration of optical performance.[7] Mehrotra[8] has pointed out that the chemistry of alkoxides is already well explored, and that a wide range of potentially useful precursors have been reported in the organometallic chemistry literature which have not yet been fully explored for sol-gel applications. For example, many heterometallic alkoxides are known whose applicability to preparations of mixed metal oxide sol-gel materials has not yet been explored in detail. Although it is necessary to consider the cost, availability, physical properties (e.g. solubility) and hydrolytic reactivity of these precursors, it seems likely that there are many useful developments to be made in this area in the future. The fact that so many studies have been restricted to a very small number of precursors highlights one of the major problems confronting every new investigator in this field, namely the very large number of potential variables (e.g. pH, R ratio, co-solvent, temperature, drying chemical control agent, aging and drying conditions etc.). Each time a new precursor is selected, in principle the effects of all these other variables may change, requiring an enormous amount of new work for systematic and complete optimisation. Despite this problem, the growing number of studies of ORMOSILS, prepared from a range of organically modified precursors, indicate the scope for precursor variation. As this area grows and new applications are found for the improved materials which result, the price and availability of the relevant important precursors can confidently be expected to become less of an obstacle.

8.6   NEW PROCESSING METHODS

Even with the existing range of precursors there is evidence of scope for future development of new processing methods. For example, acid catalysed hydrolysis at

80–100°C in a stirred sealed vial of mixtures of TMOS, MTMS and dimethyl-dimethoxysilane with a typical water:silane molar ratio of 1.4–1.8 and at least one alkyl group per silane on average, followed by vacuum evaporation of nearly half of the mixture leads to a resin which can be cast in a mould and cured to good dimensional stability within a few days. This "Fast Sol-Gel" process is useful for replication of microoptical elements.[9] Another recent useful development in sol-gel processing concerns use of supercritical CO_2 as a reaction medium to form aerogels directly.[10] Although the use of supercritical drying of conventionally-made sol-gel samples is well known, it is difficult to obtain the necessary conditions for reaction of water with alkoxides in supercritical CO_2 due to poor miscibility of water and the supercritical CO_2. However, another rapid sol-gel method[11] involving direct reaction of alkoxides with formic acid in the absence of water was found to work in supercritical CO_2 yielding gels within 12 hours and generating aerogels directly on venting the gas.[10] This method, in addition to its rapidity, opens up a wide range of new approaches to sol-gel processing as the supercritical CO_2 solvent is relatively non-polar and thus interacts in a different way with the developing gel than does the alcohol in the conventional process. This may provide further control of the structure of sol-gel materials from given precursors.

8.7 SOL-GEL SUPRAMOLECULAR CHEMISTRY

A truly ambitious ultimate future goal of sol-gel processing would be to combine the relatively gentle chemical conditions of sol-gel chemistry with an extension of supramolecular chemistry to develop self-assembling structures. Nature produces a wide range of inorganic skeletal structures in this way, and the growing understanding of biomineralisation processes suggests that it may be realistic to explore this route for development of new materials with unique properties via synthetic routes having the added advantage of high energy efficiency and specificity.[12] Because the self-assembly processes involved provide the potential to synthesise materials over a very wide range of dimensional scales from molecular through nanoparticles to the dimensions of common functional objects, this strategy has recently been described as "panoscopic".[13] The obstacles to such an approach appear formidable: the strong long-range electrostatic forces in ionic solids contrast markedly with the short-range weak forces directing molecular assembly and supramolecular chemistry, while the use of the alkoxy-precursor route in sol-gel synthesis inevitably involves the presence of water and alcohols which tend to compete in, and hence disrupt, hydrogen bonding interactions which are amongst the strongest driving forces for self-assembly. These problems have indeed resulted in failure for many attempts to use supramolecular structures for templating pores in sol-gel materials. Nevertheless, the perspective on the ultrastructure of inorganic solids provided by, for example, high resolution electron microscopy indicates that the idea of single thermo-dynamically-dominant phases defined by the deep potential minima afforded by strong cooperative interactions is far from universally applicable. Thus, some apparently non-stoichiometric minerals have been shown to contain topologically compatible coherent intergrowths of different structure and composition to the surrounding host matrix.[14] This gives a clue to successful supramolecular control of inorganic solid structures — namely the importance of surface interactions and topological compatibility in the design of successful templating processes. Thus long chain quaternary ammonium salts form micellar structures which are able to bind soluble silicate species, with the resultant units self-assembling into a liquid-crystalline array followed by condensation of the silicate groups in the array interstices to give a mesoporous silica with rod-like channels (MCM41).[15] A second important aspect is the controlled modification of the strong long-range interactions which lead to deposition of conventional crystalline inorganic solids. This is achieved in sol-gel chemistry by controlled hydrolysis and condensation processes, but also by the use of organically-modified precursors with large organic groups which are retained during hydrolysis and serve to increase the spacing between growing colloid particles and hence reduce coagulation rates. This often results in nanoparticles, whose assembly can then be controlled either by surface modification[16] or by control of the ionic strength. An elegant example of the latter is the formation of a macroporous silica material by impregnation of a pre-organised bacterial rod structure with silica nanoparticles followed by drying and thermal removal or the organic phase. During the drying the rod-structure contracts, with an associated increase in ionic strength which leads to aggregation of the nanoparticles.[17] A third and related factor is the control of the amount of inorganic material available for building a solid structure at any given time. This may be kinetic control, by limiting the hydrolysis and condensation rates, or control achieved by isolation of small quantities of material

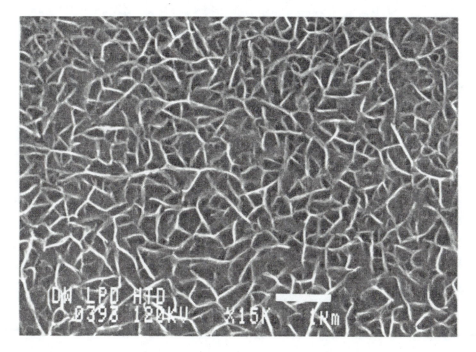

FIGURE 8.1   CELLULAR FILM OF LEPIDOCROCITE (FeOOH). (COURTESY OF PROFESSOR S. MANN).

in a given region of space — for example in a micelle or vesicle or cellular foam or microemulsion. Thus, for example, cellular films of lepidocrocite (FeOOH) have been prepared by enhanced oxidation at the interfaces of a biliquid foam formed by an oil/surfactant microemulsion with deaerated aqueous Fe(II) solution (Figure 8.1). In this example both the limited amount of metal ion and the rapid rate of surface oxidation reaction lead to a thin film of the oxide solid replicating the open cellular foam structure.[12] In relation to the history of sol-gel chemistry and even organic supramolecular chemistry, the development of sol-gel synthesis of organised matter is still its early stages and it is too early to predict with any confidence whether it will lead to practical new materials on a commercial scale. However it represents considerable and exciting challenges with the prospect of new generations of materials prepared in potentially efficient and environmentally friendly ways, and may therefore be expected to feature as a key theme of this area of chemistry in the future.

REFERENCES

1. D.R. Uhlmann and G. Teowee, *J. Sol-Gel Sci. Technol.*, **13**, 153–162 (1998).
2. G.W. Scherer, S. Calas and R. Sempéré, *J. Sol-Gel Sci. Technol.*, **13**, 937–943 (1998).
3. J.H. Strange and J.B.W. Webber, *Meas. Sci. Technology*, **8**, 555–561 (1997).
4. B. Dunn and J.I. Zink, *J. Mater. Chem.*, **1**, 903 (1991); *Chem. Mater.*, **9**, 2280–2291 (1997).
5. N.A.J.M. Sommerdijk and J.D. Wright, *J. Sol-Gel Sci. Technol.*, **13**, 565–8 (1998).
6. M.S.W. Wong, N. Bazin and P.A. Sermon, *J. Sol-Gel Sci. Technol.*, **8**, 499–505 (1997).
7. I.M. Thomas, *J. Sol-Gel Sci. Technol.*, **13**, 713–6 (1998).
8. R.C. Mehrotra, *Mater. Res. Soc. Symp. Proc.*, **121**, 81 (1988).
9. Y. Haruvy and S.E. Webber, *Chem. Mater.*, **3**, 501–7 (1991); *U.S. Pat. 5,272,240* (1993).
10. D.A. Loy, E.M. Russick, S.A. Yamanaka, B.M Baugher and K.J. Shea, *Chem. Mater.*, **9**, 2264–8 (1997).
11. K.G. Sharp, *J. Sol-Gel Sci. Technol.*, **2**, 35–41 (1994).
12. S. Mann, S.L. Burkett, S.A. Davis, C.E. Fowler, N.H. Mendelson, S.D. Sims, D. Walsh and N.T. Whilton, *Chem. Mater.*, **9**, 2300–2310 (1997).
13. G.A. Ozin, *Chem. Commun.*, 419 (2000).
14. J.S. Anderson, *Chem. Britain*, **13**, 182 (1977).
15. C.T. Kresge, M.E. Leonowicz, W.J. Roth, J.C. Vartuli and J.S. Beck, *Nature*, **359**, 710 (1992).
16. H.K. Schmidt in *"Applications of organometallic chemistry in the preparation and processing of advanced materials"*, edited by J.F. Harrod and R.M. Laine, p. 47–67. Kluwer: Dordrecht, 1995.
17. S.A. Davis, S.L. Burkett, N.H. Mendelson and S. Mann, *Nature*, **385**, 420 (1997).

INDEX

Acid catalysis, 15, 19, 34, 61

Acid dissociation of hydrated metal ions, 53–56

Actuator materials, 102

Advantages of sol–gel materials, 4–5, 63, 85, 87, 99, 101–104, 109

Aerogels, 13, 28, 31, 100–101

Ageing, 4, 24–26, 36–37

Agglomeration of dye molecules, prevention of, 87

Alcoholysis, 19

Alcoxolation of metal hydroxy species, 60

Atomic force microscopy, 81

Babel, 6

Bacteria, 40, 113

Base catalysis, 15, 19, 34, 61

BET isotherm, 73

Bidentate ligands, 33

Bioactive glasses, 45, 90

Biomineralisation, 67, 112–114

Biosensors, 90–91

Bloating, 49

Bragg scattering, 77–80

Brownian motion, 1

Buckyballs, 45

Cage compounds, 41

Capacitor materials, 102

Capillary stress, 4, 26–28, 37–38, 86

Catalysts, 5, 8–9, 22, 34, 38–39, 91–93, 99–101

Ceramic composites, 98–99

Ceramic membranes, 94

Ceramics, 7, 9–11, 64–65, 97–99

Ceramics, fibre–reinforced, 98–99

Chain length effects, alkoxide precursors, 17, 33

Chelating ligands, 62

Chemical sensors, optical, 89–90

Chemical sensors, semiconductor, 104–106

Chemical shift, 70

Cloudiness, 27, 37

Coagulation, 3

Coarsening, 24

Coatings, 11–12, 93, 103

Colloid particle size, 1–3, 45

Composites, 1, 7, 13, 38–48, 71, 83, 88–90, 93, 98, 109–110

Concrete, 7

Condensation of metal oxide species, 56–59

Condensation, 4, 16, 19–21, 33–35, 56–59

Condensation, role of catalyst, 34

Condensation, role of precursor, 33

Condensation, role of water, 35

Constant-rate period, 26

Contact angle, 74

Contact lenses, 47, 89

Coordinative unsaturation of metal alkoxides, 61

Co-solvent, 17–18, 25, 35

Counter-ion, effects on diffuse layer thickness and coagulation, 2–3, 59

Counter-ions, coordination to metal centres, 59–60

Counter-ions, preferential adsorption of, 59

Cracking, 4, 5, 26–28, 31, 34, 37–38, 42, 86

Critical point, 26

Cross polarisation technique, 70

Crystallite size determination, 80

Dehydroxylation, 48–50

Densification, 4, 28–31

Depth-profiling of composition, 81–82

Derivatisation, 43

Diffuse double layer, 2, 64–66

Dip coating, 93

DLVO theory, 65–66

Doped glasses, 38–40, 43–46, 87–89, 91–92

Doped metal oxides, 63, 103–106

Drying control chemical additive (DCCA), 18, 28, 37–38, 42

Drying, 4–5, 8, 18, 26–28, 33–34, 36–38, 42, 47–48, 77, 82, 85–87, 100–101, 110–113

Eddystone, 7

Efflorescence, 27

Elasticity, 23

Electrochromic materials, 102–103

Electron energy loss spectroscopy (EELS), 81

Electron microscopy, 81

Electronegativity, 54

Electronic materials, 101–104

Emulsions, gelation of, 66

Energy dispersive X–ray analysis (EDAX), 81

Entrapment, 43, 89–90

Entrapped analytical reagents, 89

Enzymes, 5, 45, 90

Esterification, 15

Extended X–ray absorption fine structure (EXAFS), 62, 80–81, 106, 110

Faience, 7

Falling–rate periods, 26–27

Faraday, 2

Fast sol-gel methods, 100, 111–112

Ferroelectric materials, 101–102

Fibres, 5, 13, 63, 86, 90, 98–99

Films, 5, 11–13, 24, 34, 41, 81, 85–87, 93, 101–103, 109, 113–114

Flammable gas sensors, 104–106

Flash protection, 45

Flocculation, 66

Flory/Stockmayer theory, 23

Fluidised bed catalytic reactors, 101

Fluorescence anisotropy, 82

Fluorescence polarisation, 82

Fluorescence probes for solvent polarity, 82

Fluorination, 50

Foam formation, 49, 114

Fractals, 78–79

Freeze drying, 28

Gel structure for acid and base catalysis, 20

Gel time, 16, 18, 34–36

Gelation models, 23

Gelation, 4, 22, 36

Gel-guest interactions, 110–111

Gel-point, 23

Gibbs–Thompson equation, 75–76

Glass transition temperature, 30

Glazes, 6–7

Gold particles, 45

Green body, 97–98

Growth of particles, factors influencing, 3, 45–46, 64–66, 110

Guinier equation, 78

Gyroids, 40

Hardness, 54

Hybrid materials, 46–48, 83, 88, 93

Hybrid materials, non-shrinking, 48

Hydraulic cement, 7

Hydrogen bonds, 38

Hydrogenated sugars, effect on hydrolysis and gel time, 35
Hydrolysis, 4, 15–21, 33–37, 40–41, 45–48, 50, 53–56, 60–63, 66, 71, 86–87, 93–94, 100, 105–106, 111–113
Hydrophobic effects, 17
Hydrosilylation, 43

Immobilisation, 91
Infra red, 62, 71–2
Isoelectric point 3, 16, 58, 66

Kelvin equation, 2, 24, 26, 30, 37, 74, 75
Kinetics of hydrolysis and condensation, 20–21
Knots, 40

Ladder structures, 41
Lamellar phases, 39
Langmuir isotherm, 72
Lascaux cave paintings, 6
Lasers, 87
Leaching, 90, 94
Leaching, prevention of, 43
Light scattering, 46, 77–79, 98
Limitations of sol–gel materials, 5, 86, 101
Liquid crystalline phases, 39
Luminescence, 82
Lyotropic phases, 39

MCM-41, 39, 92, 113
Membranes, 94
Mercury porosimetry, 74
Mesoporous materials, 39, 73, 91–93, 113
Metal ion catalysts in silica, 92
Metal oxide gels, 53–67
Metal oxide polyanions, 58–59
Microporous materials, 73, 99
Microscopic viscosity, 83
Minimum contrast method in neutron scattering, 79

Miscibility, 17–18, 28, 63
Molecular footprints, 29, 38–40, 91
Monoliths, drying of, 27, 42

Nanocrystalline sol-gel materials, 62, 66, 80, 98, 105–106
Nanoscale mixing, 46
Neutron scattering, 77–80
Nitrogen adsorption porosimetry, 72–74
NMR, 62
NMR imaging of pore size, 110
NMR linewidth, 70
NMR of composites, 110
NMR spin-lattice relaxation, 76
NMR spin-spin relaxation, 76
NMR, ^{17}O, 71
NMR, ^{29}Si, 21, 24, 69–71
Non–hydrolytic sol–gel processing, 21–22, 63–4
Non-linear optic (NLO) materials, 86–89
Nucleation, 3, 66–67

Olation, 57, 60
Optical materials, 85–89
Organic templates, 38–40
Organics, loss of on heating, 29
Organometallic dopants, 46
Ormosils, 41–47
Ostwald ripening, 2
Oxo species, 55–59
Oxolation, 58–60

Partial charge model, 54–56
Particle growth and aggregation, 3, 45–46, 64–66, 110
Particle size determination, 80–81, 106
Pauling electronegativities, 54
Peptization, 65
Percolation models, 23
pH, 3, 5, 15–16, 19–20, 24–25, 28, 33–36, 45, 56, 58–60, 66

Phase diagram, 18
Phase separation, 25, 35, 42, 63
Phase transformations, 25
Photochemical hole-burning, 87
Photochemistry, 44
Photochromic glasses, 44
Photochromism, 87–88
Photoluminescence, 87
Photophysics, 44
Piezoelectric materials, 102
Pore shape determination, 75–77, 110
Pore size distribution, 4, 5, 20, 22,
    25–26, 28–29, 33–34, 36–37, 43,
    74–77, 101, 110
Pore–size control, 28, 36
Porod region, 78–79
Porod slope, 78–80
Porosity, 5, 33, 42–43, 74, 85–86,
    93–94, 100–102, 109–110
Portland cement, 7
Precursor developments, 111
Purity, 4, 63, 99–101
Pyroelectric materials, 102
Pyrolysis, 30

Q^n notation, 24, 69–70

R ratio, 16, 18, 35
Radius of gyration, 78
Raman spectroscopy, 71–72
Resins, 41
Reverse saturable absorbers, 45
Rigidochromism, 82
Ripening, 24
Rutherford backscattering spectroscopy,
    82

Scanning Auger microscopy, 81
Scattering vector, 77–80
Scherer equation, 80
Sedimentation, 1
SEM, 81
Shrinkage, 4–5, 24, 26, 28–30, 42,
    46, 48, 86–87, 99, 102

Silica, helical ribbons of, 40
Silicon oxynitride fibres, 86
Sintering, 30, 64
Small–angle scattering, 77–79, 98
Smoluchowski equation, 23–24, 64,
    67
Solid electrolyte materials, 103
Soil conditioner, 3
Solubility, 3, 24–25, 36, 46–47, 111
Solvatochromic probes, 83
Sonication, 45
Sonogels, 86
Sonosols, 86
Spanning cluster, 23
Spherical particles, 1, 9–10, 35, 64,
    66–67, 79, 98
Spheroids, 40
Spin coating, 24, 93, 101–102
Steric effects, 17, 20, 33, 61–62
Stokes' Law, 1
Stress failure, 64
Substituent effects on condensation
    reactions, 20, 41
Substituent effects on hydrolysis rates,
    16–17, 41
Substrates for microelectronics, 103
Sunglasses, 103
Supercritical drying, 28, 37, 100–101,
    112
Supercritical fluids as sol–gel reaction
    medium, 112
Supramolecular assemblies as mesopore
    templates, 38–40, 91, 112–114
Surface area, changes on aging, 36
Surface area, changes on densification,
    30
Surface charge, 2–3, 65–66
Surface functionalisation, 48–50
Surface functionality, 33
Surface modification, 48–50
Surface tension, 2, 26, 38, 74, 101
Surfactants, 38–39, 113
Synchrotron radiation, 79
Syneresis, 4, 22, 24, 26

TEM, 81

Templating in catalysis, 29, 38–40, 91

Thermoporosimetry, 74

Tin oxide sensor materials, 104–106

Trans-esterification, 18

Transparency, 33, 41, 43

Triplet states, 45

Tube-like silicates, 40

Two-fraction fast-exchange model, 76

Ultrasonic mixing, 17, 45, 86

Van der Waals equation, 72

Vibrational spectroscopy, 71–72

Viscosity in densification, 30

Viscosity, 1, 3–4, 23, 30, 82–83, 87

Viscous flow, 30

Viscous sintering, 30

Water/alcohol ratio and catalyst activity, 35

Water:alkoxide ratio, 16, 18, 35

Water glass, 7, 34, 85

Waveguide gratings, 87

Weight loss, 28–30

XANES, 62

Xerogels, 28, 38, 43, 48

X–ray photoelectron spectroscopy, 81

X–ray scattering, 77–79, 98

XRD line broadening, 80–81, 106

Yiftah El, 6

CHEMICAL SUBSTANCE INDEX

γ-alumina, 94

$(C_5H_5)_2Pd(\eta^3-C_3H_5)$, 92

$(C_5H_5)_2TiCl_2$, 92

1,3-propanediol, 102

2,6-di-*tert*-butyl phenol, 92

2-ethylhexanol, 66

2-methoxyethanol, 102

2-methyl-pentane-2,4-diol, 33

3-aminopropyltrialkoxysilane, 92

3-chloropropyltrialkoxysilane, 92

3-glycidoxypropyltrimethylsilane, 47

3-glycidyloxypropyltrialkoxysilane, 92

6-propionyl-2-(dimethylamino)
     naphthalene, 83

8-hydroxy-1,3,6-pyrenesulphonic acid,
     82

Acetaldehyde, 110

Acetate, 34, 59, 100

Acetic acid, 34, 102, 110

Acetonitrile, 36, 37

Acetylacetonates, 100

Acetylacetone, 34, 105

Alkaline phosphatase, 45

Alkenes, 92

Alkyl ammonium salts, 29

Aluminium, 3

Aluminium acetylacetonate, 12

Aluminium alkoxides, 9, 10

Aluminium *sec*-butoxide, 100

Aluminium chloride, 22

Aluminium-doped silica, 92

Aluminium hydroxide, 3

Aluminium nitride, 103

Aluminium oxide, 8–11, 13, 98, 103

Aluminium titanate, 63

Aluminosilicates, 39

Aminoalkyl silanes, 43

Ammonia, 86, 110

Ammonium chloride, 34

Ammonium fluoride, 34, 50

Antimonic acid, 103

Aspartase, 45

Asphalt, 7

Barium strontium titanate, 102

Barium titanate, 102

Benzyl alcohol, 22

Benzylic groups, 22

Bis(cyclopentadiene)zirconium
     dichloro complex, 91

Bisulphate, 60

Bitumen, 6

Boehmite, 98

Boron oxide, 8, 9

Caesium fluoride, 34

Calcium, 6

Calcium fluoride, 7

Calcium ions, 3

Calcium oxide, 7, 8

Calcium sulphate, 3

Carbon, 6

C_{60}, 45

Carbon dioxide, 4, 28, 112

Carbon monoxide, 90, 105

Carbon tetrachloride, 50

Cerium oxide, 102
Cetyltrimethylammonium chloride, 39, 40
Chalk, 7
Chloride, 59
Chromium, 56
Chromium hydroxy-aquo complexes, 57
Chromium oxide, 9,12, 39
Clays, 6
Cobalt alkoxides, 62
Cobalt oxide, 8, 9
Concrete, 7
Copper, 90, 98, 106
Copper alkoxides, 62
Copper oxide, 8, 12, 39
Copper salts, 7, 8
Copper zirconate, 8
Cyclodextrin, 38, 40

Deuterium oxide, 79
Dihydrogen phosphate, 59
Dimethoxybenzene, 45
Dimethyl (hexafluoroacetylacetonato)-gold, 46
Dimethyl (trifluoroacetylacetonato)-gold, 46
Dimethyldimethoxysilane, 112
Dimethylformamide, 17, 36, 37
Di-n-propyl ether, 64
Dioxane, 17, 35, 36
DL-tartaric acid, 39, 40

Epoxyalkyl silanes, 43
Epoxysilanes, 89
Eriochrome cyanine R, 90
Ethylbenzoate, 38
Ethylene glycol monomethylether, 36
Ethylene glycol, 36
Ethyltrimethoxysilane, 88

FeOOH, 79, 114
Ferric chloride, 3, 22
Ferric oxide, 102

Ferric oxide, 3, 59
Flint, 7
Fluoride ions, 36
Formamide, 17, 28, 36, 37, 66
Formic acid, 12

Glucose oxidase, 90
Glucose, 2, 91
Gold sols, 2, 45, 98
Gypsum, 3

Haematite, 7
Hafnia, 10
Horseradish peroxidase, 90
Hydrochloric acid, 34, 35
Hydrofluoric acid, 34
Hydrofluoric acid, 50
Hydrogen peroxide, 90
Hydrogen, 94, 100, 105
Hydroxyethyl methacrylic acid, 89
Hydroxymethyl ferrocene, 91

Ice, 75
Indium sec-butoxide, 103
Indium tin oxide, 102-103
Invertase, 45
Iridium, 44
Iron, 56, 106
Iron alkoxides, 8, 62
Iron oxide, 6-9
Isopropyl chloride, 64

$La_{0.9}Sr_{0.1}Ga_{0.8}Mg_{0.2}O_{2.85}$, 103
Lactides, 92
Lead acetate, 102
Lead oxide, 12
Lead zirconium titanate, 101–102
Lepidocrocite, 114
$Li_{1.3}Al_{0.3}Ti_{1.7}(PO_4)_3$, 103
Lime, 6, 7
Lithium aluminosilicate, 103
Lithium borosilicate, 103
Lithium gallosilicate, 103
Lithium niobate, 102

Magnesium fluoride, 7
Magnesium hexafluorosilicate, 7
Magnetite, 7
Manganese, 56
Manganese alkoxides, 62
Manganese oxide, 12, 39
Mannitol, 35
Mercury, 74, 110
Merocyanines, 88
Metal halides, 12, 21
Methacrylates, 47
Methacrylic acid, 98
Methane, 105
Methyl methacrylic acid, 89
Methyl viologen, 44
Methyltriethoxysilane, 41
Methyltrimethoxysilane, 86, 112
Molybdenum oxide, 9
Molybdenum oxo-hydroxy complexes, 58
Mullite, 10

N,N-dimethylamino pyridine, 34
N,N'-tetramethylene 2,2'-bipyridinium bromide, 44
$Na_4Zr_2Si_3O_{12}$, 103
$Na_5GdSi_4O_{12}$, 103
$Na_5YSi_4O_{12}$, 103
Nasicon ($Na_{1+x}Zr_2Si_xP_{3-x}O_{12}$), 103
Nickel acetate, 100
Nickel alkoxides, 62
Nickel hydroxy-aquo complexes, 57
Nickel oxide, 8, 9, 12
Nickel/alumina aerogels, 100
Niobium pentoxide, 39, 102
Niobium tetraethoxide, 53
Nitric oxide, 90
Nitrogen, 72
N-methylimidazole, 34
Norbornene, 92
n-propyl chloride, 64

Octaphenylcyclotetrasiloxane, 70
Oils, 6

Oxalate oxidase, 90
Oxalic acid, 37
Oxazine-170, 88
Oxygen, 90

Palladium, 105–106
Perchlorate, 60
Peroxydase, 45
Phenyltriethoxysilane, 71
Phospholipids, 39
Phosphomolybdic acid, 59
Phosphonates, 38
Platinum, 100, 105
Poly(dimethylsiloxane), 47, 88
Poly(vinyl)pyridines, 46
Polyamines, 46
Polyethylene glycol, 45, 87, 94, 105, 110
Polyethylene oxide, 45, 87, 94, 105, 110
Polymethylmethacrylate, 83
Poly-p-phenylene vinylene, 88
Polysilsesquioxanes, 41, 42
Portland cement, 7
Potassium silicate, 28
PRODAN, 83
Propanol, 33
Propene, 94
Propionic acid, 98
Propylmethacrylate, 47
Pseudo-brookite (β-Al_2O_3), 63
$Pt(NH_3)_4Cl_2$, 92
Pyranine, 82
Pyrene, 44, 83

Quicklime, 7

$ReCl(CO)_3$-2,2'-bipyridine, 82
Resins, 41
Rhodamine 6G, 83
Ru-Sn-Al_2O_3, 100
Ruthenium porphyrin, 92
Ruthenium, 44, 56

Sand, 7

Seawater, 2

$Si(OC_6H_{13})_{4-n}(OC_2H_5)_n$, 17

$Si(OEt)_2[OCHMe(CH_2)_5CH_3]_2$, 17

$Si(OEt)_2[OCHMeCH_2CH(CH_3)_2]_2$, 17

Silicic acid, 8, 13

Silicon acetate, 110

Silicon carbide, 13, 99

Silicon nitride, 13

Silicon oxynitride, 86

Silicon tetraacetate, 34

Silicon tetrachloride, 8

Silsesquioxanes, 41, 42

Silver, 56, 98

Silver oxide, 12

Sodium fluoride, 34

Sodium ions, 3

Sodium oxide, 30

Sorbitol, 35

Spiropyranes, 88

Stilbene, 93

Stilbene oxide, 93

Strontium bismuth niobate, 102

Strontium bismuth tantalate, 102

Sulphur, 2

Tantalum oxide, 39

Tantalum tetraethoxide, 53

Tertiary allylic groups, 22

Tetra(2,6-dimethyl-heptyl)silane, 17

Tetrabutyl ammonium fluoride, 34

Tetrabutyloxysilane, 17

Tetrachlorosilane, 50

Tetrafluorosilane, 50

Tetrahexyloxysilane, 17

Tetrahydrofuran, 17, 35

Thionyl chloride, 50

Thorium oxide, 8, 98

$Ti(OPr^i)_2(acac)_2$ complex, 62

$Ti(OPr^i)_3acac$ complex, 62

Tin chloride, 12

Tin oxide 12, 102, 104–106

Tin *sec*-butoxide, 103

Tin t-butoxide, 105

Tin tetrachloride dihydrate, 63

Tin-doped silica, 92

Titanium, 56

Titanium alkoxides, 47, 53, 61-63, 89, 102

Titanium chloride, 22

Titanium-doped silica, 92

Titanium isopropoxide, 61-63, 102

Titanium n-butoxide, 62

Titanium nitride, 98

Titanium oxide, 9, 12, 39, 79, 92, 102

Titanium oxychloride, 12

Titanium phosphate, 103

Titanium silsesquioxane, 92

Titanium t-amyloxide, 62

Titanium tetraethoxide, 53

Triethanolamine, 102

Trimethylbenzene, 39

Trypsin, 45

Tungsten oxide, 9, 100, 102, 103

Tungsten tetraethoxide, 53

Uranium oxide, 8, 98

Vanadium, 56

Vanadium pentoxide, 39, 91, 102, 103

Vanadyl tetraethoxide, 53

Vinylalkyl silanes, 43

Yeast, 45

Yttrium manganate, 102

Zeolites, 92, 110

Zinc oxide, 12, 39

Zirconium, 56

Zirconium chloride, 22

Zirconium hydroxide, 8

Zirconium hydroxy-aquo complexes, 58

Zirconium isopropoxide, 102

Zirconium oxide, 8, 9, 10, 13, 79, 98, 100

Zirconium n-propoxide, 61
Zirconium phosphate, 103
Zirconium tetraethoxide, 53

Zirconium titanate, 63
Zirconopyrophyllite, 8
Zirconyl salts, 8